Representations of Quantum Algebras and Combinatorics of Young Tableaux

University

LECTURE

Series

Volume 26

Representations of Quantum Algebras and Combinatorics of Young Tableaux

Susumu Ariki

American Mathematical Society
Providence, Rhode Island

【 $A^{(1)}_{r-1}$ 型量子群の表現論と組み合わせ論 】

$A^{(1)}_{r-1}$-GATA KEIRYOSHIGUN NO HYOGENRON TO KUMIAWASERON
(REPRESENTATION OF QUANTUM ALGEBRAS OF TYPE $A^{(1)}_{r-1}$
AND COMBINATORICS OF YOUNG TABLEAUX)
by Susumu Ariki

Originally published in Japanese
by Sophia University, Tokyo, 2000

Translated from the Japanese and revised by the author

2000 *Mathematics Subject Classification.* Primary 05E10, 17B37, 17B67, 20C08;
Secondary 14M15, 16D90, 16G20, 20C33.

Library of Congress Cataloging-in-Publication Data

Ariki, Susumu, 1959–
Representations of quantum algebras and combinatorics of Young tableaux / Susumu Ariki ; [translated from the Japanese and revised by the author].
p. cm. — (University lecture series, ISSN 1047-3998 ; v. 26)
Includes bibliographical references and index.
ISBN 0-8218-3232-8 (acid-free paper)
1. Quantum groups. 2. Representations of groups. 3. Bases (Linear topological spaces) 4. Young tableaux. I. Title. II. University lecture series (Providence, R.I.) ; 26.

QA176.A7513 2002
530.14′3—dc21 2002025869

♾ The paper used in this book is acid-free and falls within the guidelines
established to ensure permanence and durability.
Visit the AMS home page at URL: http://www.ams.org/

10 9 8 7 6 5 4 3 2 1 07 06 05 04 03 02

Contents

Preface

Quantum groups are in fact not groups. They are also called quantized enveloping algebras, or quantum algebras for short. They were born in mathematical physics and have evolved into a vast area of research. In particular, they have found applications in algebraic groups and given rise to big progress in the Lusztig conjecture for algebraic groups. But I have another story to tell: these quantum algebras also gave rise to new combinatorial objects and have influenced the combinatorics related to representation theories. This research area is called "combinatorial representation theory".

These lecture notes are based on my lectures delivered at Sophia university in 1997, and are intended for graduate students who have interests in this area.

In the preparation of the lectures, I benefitted from two important papers, [**Kashiwara**] and [**Lusztig**]. In fact, my primary intention was to introduce the reader to the theory of crystal bases and canonical bases by working out special examples, quantum algebras of type $A^{(1)}_{r-1}$.

In the lectures, I have named fundamental theorems about crystal bases and canonical bases of quantum algebras as the first, the second and the third main theorems of Kashiwara and Lusztig. I hope that the naming is accepted by the society of mathematicians.

The plan of the book is as follows. The first three chapters are a preparation to start running. In the 4th to the 6th chapters, we establish basic notions of crystal bases. We then introduce canonical bases in Chapter 7 and prove fundamental theorems in the subsequent two chapters. These chapters have flavors of the general theory, although we are content with our examples, the quantum algebras of type $A^{(1)}_{r-1}$. In the next two chapters, we turn to combinatorics. We prove the combinatorial construction of crystal bases of Fock spaces due to Misra and Miwa. In the 12th chapter, we summarize its applications to the representation theory of cyclotomic Hecke algebras. The 13th and 14th chapters are devoted to the proof of my main theorem stated in Chapter 12. The final chapter is a guide for further reading. The list is not intended to be complete of course, and reflects my personal research interests.

If the reader has some familiarity with representation theory, I recommend skipping the first three chapters. If he/she has some specialty in this field, I recommend starting with the 7th chapter.

I would like to thank Professor Bhama Srinivasan and Andrew Mathas for reading the manuscript, and my wife Tomoko for many things. During the preparation of these notes, I was partially supported by the JSPS-DFG Japanese-German cooperative science program "Representation Theory of Finite and Algebraic Groups".

Susumu Ariki

CHAPTER 1

Introduction

1.1. How do you do ?

These lectures are for graduate students who know the basics of the representation theory of finite groups and artinian rings. Through these lectures, you will be exposed to some recent research in mathematics. Since we have to skip the proofs of several theorems at several points to keep these lectures elementary, you are encouraged to read the original papers for these parts in the second reading. In the first reading, I recommend trusting the results so as to make your life easy. Although I skip proofs at several points, these notes are basically written in the "theorem and proof "style.

The main example we use is the quantum algebra of type $A^{(1)}_{r-1}$. The purpose of the first half of this book is to explain the general theory of crystal bases using this example. In the second half, we explain several interesting results using Young diagrams. I hope that the reader finds it interesting to do research in "Combinatorial Representation Theory"after reading these notes.

1.2. What are we interested in ?

When you start your professional education in mathematics, you soon encounter the notion of a group. It is a mathematical device used to describe symmetries in nature. It is an idea that developed concurrently in several areas, such as geometry, number theory and the algebra of polynomial equations, with an axiomatic definition coming in the middle of the 19th century. As you already know, the Galois theory is the most famous application of the group theory. In modern times groups are widely used in many areas and they have little to do with equations. They are also used in essential ways in physics and chemistry. For example, the groups used in gauge theories and classification of elementary particles are called **Lie groups**. A typical example of a Lie group is the matrix group

$$GL(n,\mathbb{C}) = \{\, X \in M(n,n,\mathbb{C}) \mid \det(X) \neq 0 \,\},$$

but there are other examples as well. Let us consider a group G. Often G will be described as a group of matrices; there are many different ways of doing this, usually using matrices of different sizes, although the group behind them all is the same. Namely, there are many ways to associate a matrix $\rho(X)$ with $X \in G$ in such a way that the product of two elements corresponds to the product of the associated matrices (i.e. $\rho(X)\,\rho(Y) = \rho(XY)$). These ρ are called **representations** of G. Lie himself worked with "infinitesimal groups", understanding that for many things it was usually enough to consider the **Lie algebras**. In today's language, we can say that an essential feature of Lie's work is his discovery that by looking at the second term of the exponent in the Campbell-Baker-Hausdorff formula

$$exp(X)exp(Y) = exp(X + Y + \frac{1}{2}(XY - YX) + \cdots),$$

we can recover information on the group structure of the Lie group. (This is no longer true if we treat matrix groups over fields of positive characteristic. Here, we consider Lie groups over $\mathbb{C}$ only.) Since we do not treat Lie groups in these lectures, we start with Lie algebras.

DEFINITION 1.1. Let $\mathfrak{g}$ be a vector space over $\mathbb{C}$ equipped with a $\mathbb{C}$-bilinear map, called the Lie bracket, $[\ ,\ ] : \mathfrak{g} \otimes \mathfrak{g} \to \mathfrak{g}$ satisfying:

$$(1)\quad [Y, X] + [X, Y] = 0,$$
$$(2)\quad [X, [Y, Z]] + [Y, [Z, X]] + [Z, [X, Y]] = 0.$$

Then $\mathfrak{g}$ is called a **Lie algebra** over $\mathbb{C}$. The condition (2) is called the **Jacobi identity**.

A linear map ϕ between two Lie algebras is called a **Lie algebra homomorphism** if $\phi([X, Y]) = [\phi(X), \phi(Y)]$ for $X, Y \in \mathfrak{g}$.

Let V be a vector space over $\mathbb{C}$. Then $\mathrm{End}(V)$ becomes a Lie algebra via $[X, Y] := XY - YX$. We denote this Lie algebra by $\mathfrak{gl}(V)$. If a basis of V is chosen and V is identified with $\mathbb{C}^n$, it is also denoted by $\mathfrak{g}l(n, \mathbb{C})$. The following definition is also important.

DEFINITION 1.2. A subspace $\mathfrak{a}$ of a Lie algebra $\mathfrak{g}$ is called a **Lie subalgebra** of $\mathfrak{g}$ if the condition $[\mathfrak{a}, \mathfrak{a}] \subset \mathfrak{a}$ is satisfied. If $[\mathfrak{a}, \mathfrak{a}] = 0$ holds, $\mathfrak{a}$ is always a Lie subalgebra. In this case, $\mathfrak{a}$ is called a **commutative** Lie subalgebra.

If $\mathfrak{a}$ satisfies a stronger condition $[\mathfrak{g}, \mathfrak{a}] \subset \mathfrak{a}$, we call $\mathfrak{a}$ an **ideal** of $\mathfrak{g}$.

The notion of representations for Lie algebras is defined in the following way.

DEFINITION 1.3. Let V be a (not necessarily finite dimensional) vector space over $\mathbb{C}$. If a $\mathbb{C}$-linear map $\rho : \mathfrak{g} \to \mathrm{End}(V)$ satisfies

$$\rho([X, Y]) = \rho(X)\rho(Y) - \rho(Y)\rho(X),$$

then ρ is called a **representation** of $\mathfrak{g}$, and V is called a **$\mathfrak{g}$-module**.

In other words, if $\rho : \mathfrak{g} \to \mathfrak{g}l(V)$ is a Lie algebra homomorphism, ρ is called a representation of $\mathfrak{g}$. If one would like to record both ρ and V explicitly, we make these into a pair (ρ, V), and call it a representation.

An important example is the **adjoint representation** $(ad, \mathfrak{g})$, where $ad : \mathfrak{g} \to \mathrm{End}(\mathfrak{g})$ is defined by $ad(X)(Y) = [X, Y]$.

Although Lie algebras appear to be strange algebras, they are in fact connected with usual (associative) algebras, and we can interpret results about a Lie algebra as results about an algebra in the usual sense; this algebra is the **enveloping algebra** of $\mathfrak{g}$, whose precise definition is given later. In Lie theory, it has become more popular to work with enveloping algebras.

Another important work relevant to us is a result of Serre. His result is that we can give a definition of enveloping algebras in terms of generators and relations for a good class of Lie algebras, semisimple Lie algebras. These relations are now called the **Serre relations**.

As research developed this way, an interesting breakthrough was made in the 1980's by leading mathematicians in the Kyoto school. They discovered that if we deform the structure constants of the Serre relations by introducing a parameter, the resulting algebras manifest hidden symmetries of certain spin models, and we can compute physically important quantities of the models. These algebras are the quantum algebras of Drinfeld and Jimbo.

In this brief history, we can see a natural evolution of mathematical ideas and the discovery of new mathematical concepts. The quantum algebra so discovered is the subject of these lectures.

1.3. Enveloping algebras

We define the enveloping algebra of a Lie algebra. The word "algebra" always means a unital associative $\mathbb{C}$-algebra in this chapter.

DEFINITION 1.4. Let $\mathfrak{g}$ be a Lie algebra. Among pairs (A, ρ) of an algebra A and a $\mathbb{C}$-linear map $\rho : \mathfrak{g} \to A$ satisfying

$$\rho([X, Y]) = \rho(X)\rho(Y) - \rho(Y)\rho(X),$$

we consider the **universal** one: namely, the pair $(U(\mathfrak{g}), \iota)$ having the property that

> for any pair (A, ρ), there exists a unique algebra homomorphism $\phi : U(\mathfrak{g}) \to A$ which satisfies $\rho = \phi \circ \iota$.

We call the algebra $U(\mathfrak{g})$ the (universal) **enveloping algebra** of $\mathfrak{g}$.

Perhaps we should denote the enveloping algebra by $(U(\mathfrak{g}), \iota)$, but it is convention to drop ι. This is harmless since, if U is isomorphic to $U(\mathfrak{g})$ as an algebra, we can define a map ι to make (U, ι) a universal pair in the above sense. It is important to note that we consider not only finite dimensional algebras but also infinite dimensional algebras A.

The universality of $U(\mathfrak{g})$ is usually visualized by a commutative diagram:

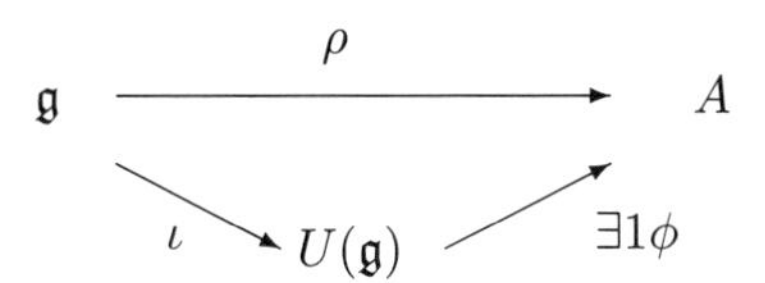

Since it is not at all clear from the definition that $U(\mathfrak{g})$ exists, it is necessary to show its existence.

LEMMA 1.5. *The enveloping algebra of a Lie algebra always exists. It is unique up to an isomorphism of algebras.*

This is a standard fact, and we leave it to the reader in the exercises below. Since the second exercise is a bit difficult for beginners, I recommend consulting a textbook on Lie algebras.

EXERCISE 1.6. Show that there is at most one enveloping algebra up to an isomorphism of algebras.

EXERCISE 1.7. Let $T(\mathfrak{g}) = \oplus_{n \geq 0} \mathfrak{g}^{\otimes n}$ be the tensor algebra over $\mathfrak{g}$, and let I be the two-sided ideal generated by

$$X \otimes Y - Y \otimes X - [X, Y] \quad (X, Y \in \mathfrak{g}).$$

Show that $U(\mathfrak{g}) \simeq T(\mathfrak{g})/I$, and $\iota : \mathfrak{g} \to U(\mathfrak{g})$ is injective. Here, ι is the map naturally induced by the inclusion $\mathfrak{g} \subset T(\mathfrak{g})$.

The way of defining an algebra as in Exercise1.7 is called **definition by generators and relations**. Since we will meet this kind of definition repeatedly in later sections, I shall give its precise definition here. The field $\mathbb{C}$ can be replaced by any commutative ring.

DEFINITION 1.8. We take a $\mathbb{C}$-vector space V and its basis $X_1, \dots, X_N$. For elements $R_1, \dots, R_M \in T(V) = \oplus_{n \geq 0} V^{\otimes n}$, we set $I = \sum_{i=1}^{M} T(V) R_i T(V)$, which is the minimal two-sided ideal of $T(V)$ containing $R_1, \dots, R_M$. The algebra $T(V)/I$ is the algebra defined by generators $X_1, \dots, X_N$ and relations $R_1 = 0, \dots, R_M = 0$.

In the above exercise, we let $X_1, \dots, X_N$ be a basis of $\mathfrak{g}$, and let c_{ij}^k be complex numbers defined by $[X_i, X_j] = \sum_{k=1}^{N} c_{ij}^k X_k$. Then the aim of the exercise is to show that the algebra defined by generators $X_1, \dots, X_N$ and relations $X_i X_j - X_j X_i = \sum_{k=1}^{N} c_{ij}^k X_k$ $(1 \leq i < j \leq N)$ is isomorphic to the enveloping algebra of $\mathfrak{g}$. Here, $X_{i_1} \cdots X_{i_N}$ stands for $X_{i_1} \otimes \cdots \otimes X_{i_N}$.

Let $\rho : \mathfrak{g} \to \mathrm{End}(V)$ be a representation of $\mathfrak{g}$. By the universality property of the enveloping algebra, there exists a unique algebra homomorphism $\phi : U(\mathfrak{g}) \to \mathrm{End}(V)$ satisfying $\phi \circ \iota = \rho$. This implies that we can extend the action of $\mathfrak{g}$ on V to the action of $U(\mathfrak{g})$ in a unique way. Using this correspondence (which associates a representation (ϕ, V) of $U(\mathfrak{g})$ with a representation (ρ, V) of $\mathfrak{g}$), we can interpret results about representations of a Lie algebra as results about representations of the corresponding enveloping algebra.

PROPOSITION 1.9. *Let V_1, V_2 be $\mathfrak{g}$-modules, $f : V_1 \to V_2$ be a $\mathfrak{g}$-module homomorphism, i.e. a linear map for which $Xf(m) = f(Xm)$ $(X \in \mathfrak{g}, m \in V_1)$ holds. Then f is a $U(\mathfrak{g})$-module homomorphism, i.e. $af(m) = f(am)$ holds for $a \in U(\mathfrak{g})$ and $m \in V_1$.*

PROOF. Let $\rho_i : \mathfrak{g} \to \mathrm{End}(V_i)$ be representations on the spaces V_i, and let ϕ_i be their unique extensions $\phi_i : U(\mathfrak{g}) \to \mathrm{End}(V_i)$ $(i = 1, 2)$. We set $W = V_1 \oplus V_2$ and define a linear operator $A \in \mathrm{End}(W)$ by

$$A(v_1 \oplus v_2) = v_1 \oplus (f(v_1) + v_2).$$

In matrix form, A is given by $\begin{pmatrix} I & 0 \\ f & I \end{pmatrix}$.

Let $\rho = \rho_1 \oplus \rho_2 : \mathfrak{g} \to \mathrm{End}(W)$. Then

$$\begin{aligned} A^{-1} \rho(X) A &= \begin{pmatrix} I & 0 \\ -f & I \end{pmatrix} \begin{pmatrix} \rho_1(X) & 0 \\ 0 & \rho_2(X) \end{pmatrix} \begin{pmatrix} I & 0 \\ f & I \end{pmatrix} \\ &= \begin{pmatrix} \rho_1(X) & 0 \\ -f \circ \rho_1(X) + \rho_2(X) \circ f & \rho_2(X) \end{pmatrix} = \rho(X). \end{aligned}$$

By the uniqueness property of ϕ, we have $A^{-1} \phi(a) A = \phi(a)$ for $\phi = \phi_1 \oplus \phi_2$. By expanding the equation $\phi(a) A = A \phi(a)$, we obtain $f \circ \phi_1(a) = \phi_2(a) \circ f$ $(a \in U(\mathfrak{g}))$, which implies $af(m) = f(am)$ for $m \in V_1$. □

As a corollary, we have the following proposition. By definition, the objects of the category $\mathfrak{g} - Mod$ are representations of a Lie algebra $\mathfrak{g}$ and the morphisms of the category are $\mathfrak{g}$-module homomorphisms.

PROPOSITION 1.10. *Let $\mathfrak{g} - Mod$ be the category of representations of a Lie algebra $\mathfrak{g}$, and let $U(\mathfrak{g}) - Mod$ be the category of representations of the enveloping algebra $U(\mathfrak{g})$. We define functors $\mathcal{F}$, $\mathcal{G}$ as follows.*

$$\begin{array}{rcl} \mathcal{F} : \mathfrak{g} - Mod & \longrightarrow & U(\mathfrak{g}) - Mod \\ (\rho, V) & \mapsto & (\phi, V) \\ f \in \mathrm{Hom}_{\mathfrak{g}}(V_1, V_2) & \mapsto & f \in \mathrm{Hom}_{U(\mathfrak{g})}(V_1, V_2) \end{array}$$

$$\begin{array}{rcl} \mathcal{G} : U(\mathfrak{g}) - Mod & \longrightarrow & \mathfrak{g} - Mod \\ (\phi, V) & \mapsto & (\phi \circ \iota, V) \\ f \in \mathrm{Hom}_{U(\mathfrak{g})}(V_1, V_2) & \mapsto & f \in \mathrm{Hom}_{\mathfrak{g}}(V_1, V_2) \end{array}$$

Then $\mathcal{F}$, $\mathcal{G}$ give isomorphisms of categories.

By virtue of the proposition, we have

- V is a submodule of W as a $\mathfrak{g}$-module if and only if V is a submodule of W as a $U(\mathfrak{g})$-module.
- V is a quotient module of W as a $\mathfrak{g}$-module if and only if V is a quotient module of W as a $U(\mathfrak{g})$-module.
- $0 \to U \to V \to W \to 0$ is an exact sequence of $\mathfrak{g}$-modules if and only if it is an exact sequence of $U(\mathfrak{g})$-modules.

We have another important operation in $\mathfrak{g} - Mod$.

DEFINITION 1.11. Let (ρ_1, V_1), (ρ_2, V_2) be representations of $\mathfrak{g}$. Then we can make $V_1 \otimes V_2$ into a representation $(\rho_1 \otimes \rho_2, V_1 \otimes V_2)$ of $\mathfrak{g}$ via

$$(\rho_1 \otimes \rho_2)(X) = \rho_1(X) \otimes 1 + 1 \otimes \rho_2(X).$$

This representation is called the **tensor product** representation of V_1 and V_2.

If we consider the extension of the tensor product representation to a representation of $U(\mathfrak{g})$, it is the tensor product representation of $U(\mathfrak{g})$ in the usual sense. That is, if we denote by ϕ_i $(i = 1, 2)$ the representations of $U(\mathfrak{g})$ which are extensions of ρ_i respectively, then the extension of $\rho_1 \otimes \rho_2$ is $\phi_1 \otimes \phi_2$.

To explain this, we start with the warning that not all categories $A - mod$ have tensor product representations. To have this operation, the algebra A needs to be equipped with an algebra homomorphism $\Delta : A \to A \otimes A$. In these cases, we may consider the tensor product $M_1 \otimes M_2$ of two A-modules M_1, M_2 as an A-module via $a \cdot (m_1 \otimes m_2) = \Delta(a)(m_1 \otimes m_2)$.

In the case of the enveloping algebra, the map Δ defined by the following proposition induces the tensor product representations.

PROPOSITION 1.12. *There exists a unique algebra homomorphism*

$$\Delta : U(\mathfrak{g}) \to U(\mathfrak{g}) \otimes U(\mathfrak{g})$$

satisfying

$$\Delta \circ \iota(X) = \iota(X) \otimes 1 + 1 \otimes \iota(X) \quad (X \in \mathfrak{g}),$$

such that for any tensor product representation $\rho_1 \otimes 1 + 1 \otimes \rho_2$ of $\mathfrak{g}$, its extension to $U(\mathfrak{g})$ is given by $(\phi_1 \otimes \phi_2) \circ \Delta$. In other words, we have the following commutative diagram.

$$\begin{array}{ccc} \mathfrak{g} & \xrightarrow{\rho_1 \otimes 1 + 1 \otimes \rho_2} & \mathrm{End}(V_1 \otimes V_2) \\ & {}_{\iota}\searrow \quad U(\mathfrak{g}) \quad \nearrow_{(\phi_1 \otimes \phi_2) \circ \Delta} & \end{array}$$

We call Δ *the* **coproduct** *of* $U(\mathfrak{g})$.

PROOF. We set $A = U(\mathfrak{g}) \otimes U(\mathfrak{g})$ in the commutative diagram to define the universality of $U(\mathfrak{g})$, and consider $\mathfrak{g} \to A$ given by $X \mapsto \iota(X) \otimes 1 + 1 \otimes \iota(X)$. Then the existence and uniqueness of Δ follows from the universality of $U(\mathfrak{g})$ and the fact that $\iota([X, Y]) \otimes 1 + 1 \otimes \iota([X, Y])$ equals

$$[\iota(X) \otimes 1 + 1 \otimes \iota(X),\, \iota(Y) \otimes 1 + 1 \otimes \iota(Y)]\,.$$

Next set $A = \mathrm{End}(V_1 \otimes V_2)$. Then we can check that the required map ϕ for the tensor product representation is given by $(\phi_1 \otimes \phi_2) \circ \Delta$. In fact, the commutativity of the diagram is verified by

$$\begin{aligned}(\phi_1 \otimes \phi_2) \circ \Delta \circ \iota(X) &= (\phi_1 \otimes \phi_2)\,(\iota(X) \otimes 1 + 1 \otimes \iota(X)) \\ &= \phi_1 \circ \iota(X) \otimes 1 + 1 \otimes \phi_2 \circ \iota(X) \\ &= \rho_1(X) \otimes 1 + 1 \otimes \rho_2(X) = \rho(X).\end{aligned}$$

Hence the uniqueness of ϕ implies that $\phi_1 \otimes \phi_2$ is the map for the tensor product representation $\rho_1 \otimes \rho_2$. □

For enveloping algebras, the following result, the PBW(Poincare-Birkhoff-Witt) theorem, is fundamental.

PROPOSITION 1.13. *If* $\{X_i\}_{i \in I}$ *is a basis of a Lie algebra* $\mathfrak{g}$*, then the following set is a basis of* $U(\mathfrak{g})$.

$$\{\, X_{i_1} \cdots X_{i_m} \,|\, i_1 \le \cdots \le i_m, m = 0, 1, \ldots \}$$

PROOF. We first show that $U(\mathfrak{g})$ is spanned by these elements. If we consider those elements $X_{i_1} \cdots X_{i_m}$ whose indices $i_1, \ldots, i_m$ are not necessarily non-decreasing, it is obvious that these span $U(\mathfrak{g})$. Choose the minimal index among $i_1, \ldots, i_m$, and move this to the left end using the relation $X_j X_k = X_k X_j + [X_j, X_k]$. Next choose the minimal index among the remaining indices and move this to the second to the leftmost position. Continue this procedure to reorder the indices $i_1, \ldots, i_m$ in non-decreasing order. Since newly appearing terms have smaller length, we apply the same procedure to these terms and after a finite number of steps, we can rewrite the original monomial as a linear combination of monomials with non-decreasing indices.

Next we show that these monomials are linearly independent. To do this, we consider indeterminates $\{z_i\}_{i \in I}$ which are in bijection with $\{X_i\}_{i \in I}$, and denote the polynomial ring generated by these indeterminates by S.

Since $\{\, z_{i_1} \cdots z_{i_m} \,|\, i_1 \le \cdots \le i_m\}$ is a basis of S, we can define an action of $\mathfrak{g}$ on S as follows. (If $m = 0$, we set $X_i 1 = z_i$.)

$$X_i z_{i_1} \cdots z_{i_m} = \begin{cases} z_i z_{i_1} \cdots z_{i_m} & (i \le i_1) \\ X_{i_1}\,(X_i z_{i_2} \cdots z_{i_m}) + [\,X_i,\, X_{i_1}] z_{i_2} \cdots z_{i_m} & (i > i_1) \end{cases}$$

Assertion 1 *These operators X_i on S are well-defined.*

We can show that $X_i z_{i_1} \cdots z_{i_m}$ is the sum of $z_i z_{i_1} \cdots z_{i_m}$ and a polynomial of degree equal or less than m by induction on m, from which the well-definedness follows.

Assertion 2 *These operators make S into a $\mathfrak{g}$-module.*

We show $X_j\left(X_k z_{i_1} \cdots z_{i_m}\right) - X_k\left(X_j z_{i_1} \cdots z_{i_m}\right) = [X_j, X_k] z_{i_1} \cdots z_{i_m}$ by induction on m. Assume that this holds up to $m-1$. We first show the formula for the cases $j \leq i_1$ and $k \leq i_1$. We may assume $j < k$ without loss of generality. If $j \leq i_1$, then the definition of $X_k z_j z_{i_1} \cdots z_{i_m}$ implies the formula. If $k \leq i_1$, then we are in the case $j \leq i_1$ and the formula follows.

Next we show the formula for the case $j, k > i_1$. We do not assume $j < k$ here. We abbreviate $z_{i_2} \cdots z_{i_m}$ by z_J. Let us start with the equation

$$X_j\left(X_k z_{i_1} \cdots z_{i_m}\right) = X_j\left(X_k z_{i_1} z_J\right) = X_j\left(X_{i_1}\left(X_k z_J\right) + [X_k, X_{i_1}] z_J\right).$$

Note that $X_k z_J$ is the sum of $z_k z_J$ and a polynomial of degree less than m. If we consider an element $X_j X_{i_1} z_k z_J$, we are in the case $i_1 \not< j, k$ and thus we have $X_j X_{i_1} z_k z_J = X_{i_1} X_j z_k z_J + [X_j, X_{i_1}] z_k z_J$. This implies that if we apply $X_j X_{i_1}$ and $X_{i_1} X_j + [X_j, X_{i_1}]$ to $z_k z_J$, we have the same element. The same is true if we apply $X_j X_{i_1}$ and $X_{i_1} X_j + [X_j, X_{i_1}]$ to a polynomial of degree less than m by the induction hypothesis. Hence, $X_j X_{i_1} X_k z_J$ equals $X_{i_1} X_j X_k z_J + [X_j, X_{i_1}] X_k z_J$. We also have that $X_j [X_k, X_{i_1}] z_J$ equals $[X_k, X_{i_1}] X_j z_J + [X_j, [X_k, X_{i_1}]] z_J$ by the induction hypothesis. To conclude, $X_j\left(X_k z_{i_1} \cdots z_{i_m}\right)$ equals

$$X_{i_1} X_j X_k z_J + [X_j, X_{i_1}] X_k z_J + [X_k, X_{i_1}] X_j z_J + [X_j, [X_k, X_{i_1}]] z_J.$$

We obtain a similar formula for $X_k\left(X_j z_{i_1} \cdots z_{i_m}\right)$. By subtracting this from the above, and using the Jacobi identity, we get

$$\begin{aligned}
&X_j\left(X_k z_{i_1} \cdots z_{i_m}\right) - X_k\left(X_j z_{i_1} \cdots z_{i_m}\right) \\
&\quad = X_{i_1}[X_j, X_k] z_J + [X_j, [X_k, X_{i_1}]] z_J - [X_k, [X_j, X_{i_1}]] z_J \\
&\quad = X_{i_1}[X_j, X_k] z_J + [[X_j, X_k], X_{i_1}]\, z_J.
\end{aligned}$$

By the induction hypothesis, this equals $[X_j, X_k]\left(X_{i_1} z_J\right)$, which is the same as $[X_j, X_k] z_{i_1} \cdots z_{i_m}$. Hence the result follows.

By Assertion 2 and the universal property of the enveloping algebra, S is a $U(\mathfrak{g})$-module. Further, if we apply $X_{i_1} \cdots X_{i_m}$ $(i_1 \leq \cdots \leq i_m)$ to $1 \in S$, we get $z_{i_1} \cdots z_{i_m}$. Since these are linearly independent elements, $\{ X_{i_1} \cdots X_{i_m} \mid i_1 \leq \cdots \leq i_m \}$ are linearly independent. □

CHAPTER 2

The Serre relations

2.1. The Serre relations

If we take a semisimple Lie algebra as $\mathfrak{g}$, we have a more concise presentation of $U(\mathfrak{g})$. (Since we only consider specific examples in these lectures, it is not necessary to know what semisimple Lie algebras are.) This implies that we do not have to deform $\mathfrak{g}$ itself to obtain a deformation of $U(\mathfrak{g})$, which is crucial in defining quantum algebras. In this chapter, we pick up the simplest example of a semisimple Lie algebra and verify Serre's presentation by a concrete argument.

We first introduce the notion of a Cartan subalgebra and a root system of this Lie algebra, and then go on to the statement and the proof of the main theorem of this chapter. As in Chapter 1, $\mathfrak{gl}(n,\mathbb{C})$ is the set of $n \times n$ complex matrices which is viewed as a Lie algebra via $[X, Y] = XY - YX$, and ad is the adjoint representation.

DEFINITION 2.1. The **special linear Lie algebra** is the Lie algebra

$$\mathfrak{sl}(n,\mathbb{C}) = \{X \in \mathfrak{gl}(n,\mathbb{C}) |\ tr(X) = 0\}.$$

We denote matrix units by E_{ij}: namely, the unique non-zero entry of E_{ij} is 1 in the (i,j)th entry. Set

$$\mathfrak{h} = \left\{ X = \sum_{i=1}^{n} c_i E_{ii} \;\middle|\; \sum_{i=1}^{n} c_i = 0 \right\}$$

and define $x_i : \mathfrak{h} \to \mathbb{C}$ by $x_i(X) = c_i$.

Then $\mathfrak{h}$ is a Lie subalgebra of $\mathfrak{sl}(n,\mathbb{C})$. It is a **Cartan subalgebra** of $\mathfrak{sl}(n,\mathbb{C})$. (All other Cartan subalgebras of $\mathfrak{sl}(n,\mathbb{C})$ are of the form $g^{-1}\mathfrak{h}g$, for some $g \in GL(n,\mathbb{C})$.) The Lie algebra $\mathfrak{sl}(n,\mathbb{C})$ admits a simultaneous eigenspace decomposition with respect to $ad(\mathfrak{h})$ as follows.

$$\mathfrak{sl}(n,\mathbb{C}) = \mathfrak{h} \bigoplus \left(\bigoplus_{i \neq j} \mathbb{C}E_{ij} \right)$$

This decomposition is called the **root space decomposition** of $\mathfrak{sl}(n,\mathbb{C})$. The simultaneous eigenvalues are 0 on $\mathfrak{h}$, and $x_i - x_j$ on E_{ij}. The latter non-zero simultaneous eigenvalues are called **roots**, and the set $\Phi = \{x_i - x_j\}_{i \neq j}$ is called the **root system** of type A_{n-1}. The roots $\alpha_i = x_i - x_{i+1}\,(1 \leq i < n)$ are called **simple roots**.

THEOREM 2.2. *Let $\mathfrak{g} = \mathfrak{sl}(n,\mathbb{C})$. Then $U(\mathfrak{g})$ is isomorphic to the $\mathbb{C}$-algebra U defined by the following generators and relations.*

Generators: $\quad e_i, f_i, h_i \quad (1 \leq i \leq n-1)$.

Relations:

$$e_i f_j - f_j e_i = \delta_{ij} h_i, \quad h_i h_j = h_j h_i,$$

$$h_i e_j - e_j h_i = \begin{cases} 2e_j & (i = j) \\ -e_j & (i - j = \pm 1) \\ 0 & (otherwise) \end{cases},$$

$$h_i f_j - f_j h_i = \begin{cases} -2f_j & (i = j) \\ f_j & (i - j = \pm 1) \\ 0 & (otherwise) \end{cases},$$

$$e_i^2 e_j - 2e_i e_j e_i + e_j e_i^2 = 0 \; (i - j = \pm 1), \quad e_i e_j = e_j e_i \; (otherwise),$$
$$f_i^2 f_j - 2f_i f_j f_i + f_j f_i^2 = 0 \; (i - j = \pm 1), \quad f_i f_j = f_j f_i \; (otherwise).$$

PROOF. We define elements $L_{ij} \in U$ $(i \neq j)$ by induction on $|j - i|$:

$$L_{i,i+1} = e_i, \quad L_{ij} = [L_{ik}, L_{kj}] \quad (i < k < j),$$
$$L_{i+1,i} = f_i, \quad L_{ij} = [L_{ik}, L_{kj}] \quad (i > k > j).$$

For $a, b \in U$, $[a, b]$ is $ab - ba$ by definition.

Assertion 1 *The L_{ij} are well-defined.*

We prove this by induction on $N = |j - i|$. Since the proof for $i > j$ is the same as the proof for $i < j$, we prove it only for $i < j$. Assume that it is already proved up to $N - 1$. We shall show for $1 \leq l \leq N - 2$ that

$$[L_{i,i+l+1}, L_{i+l+1,i+N}] = [L_{i,i+l}, L_{i+l,i+N}]. \tag{2.1}$$

Since we have $L_{i,i+l+1} = [L_{i,i+l}, e_{i+l}]$ by the induction hypothesis, the left hand side of (2.1) equals $[[L_{i,i+l}, e_{i+l}], L_{i+l+1,i+N}]$, which is the same as

$$-[e_{i+l}, [L_{i,i+l}, L_{i+l+1,i+N}]] + [L_{i,i+l}, [e_{i+l}, L_{i+l+1,i+N}]].$$

We now notice that $L_{i,i+l}$ and $L_{i+l+1,i+N}$ are non-commutative polynomials in $e_i, \ldots, e_{i+l-1}$ and $e_{i+l+1}, \ldots, e_{i+N-1}$ respectively. Hence the first term is 0. The second term equals the right hand side of (2.1) by the induction hypothesis.

Assertion 2 *The Lie algebra $\mathfrak{g}$ is an irreducible U-module via the following action:*

$$e_i \mapsto ad(E_{i,i+1}), \quad f_i \mapsto ad(E_{i+1,i}), \quad h_i \mapsto ad(E_{ii} - E_{i+1,i+1}),$$

for $1 \leq i \leq n - 1$. (Recall that the E_{ij} are matrix units.)

To prove that this defines an action of U, we check that $ad(E_{i,i+1})$, $ad(E_{i+1,i})$ and $ad(E_{ii} - E_{i+1,i+1})$, for $1 \leq i \leq n-1$, satisfy the defining relations of U. By the Jacobi identity, it is enough to see that $E_{i,i+1}, E_{i+1,i}$ and $E_{ii} - E_{i+1,i+1}$ satisfy the relations, but this is easily checked by direct computation. We remark here that we also have seen that $L_{ij} \mapsto ad(E_{ij})$.

Next we show that $\mathfrak{g}$ is irreducible as a U-module. Let $\mathfrak{a}$ be a non-zero U-submodule of $\mathfrak{g}$. It is easy to see that $\mathfrak{a}$ is not contained in $\mathfrak{h}$ and that $\mathfrak{a}$ has a simultaneous eigenspace decomposition with respect to $ad(\mathfrak{h})$. So $\mathfrak{a}$ contains a matrix unit E_{kl} $(k \neq l)$. We shall show that $\mathfrak{a}$ contains all E_{ij} $(i \neq j)$. If $i \neq k$ and $j = l$, then $E_{ij} = ad(E_{ik})E_{kl} \in \mathfrak{a}$. If $i = k$ and $j \neq l$, then $E_{ij} = -ad(E_{lj})E_{kl} \in \mathfrak{a}$. If $i \neq k$ and $j \neq l$, we consider $ad(E_{ik})ad(E_{lj})E_{kl}$ and we have that E_{ij} is in the submodule $\mathfrak{a}$. Since all off-diagonal matrix units are in the submodule, we have that $ad(E_{i,i+1})E_{i+1,i} \in \mathfrak{a}$ for all i. We have proved that any non-zero submodule must coincide with $\mathfrak{g}$.

Assertion 3

(1) The algebra U is a U-module via $ad(x)(a) = xa - ax$ $(x = e_i, f_i, h_i)$.
(2) The space $L = \sum_{i=1}^{n-1} \mathbb{C}h_i + \sum_{i \neq j} \mathbb{C}L_{ij}$ is stable under $ad(L)$. In particular, it is a U-submodule.
(3) Define a linear map $\iota : \mathfrak{g} \to L$ by

$$E_{ij} \mapsto L_{ij} \ (i \neq j), \quad E_{ii} - E_{i+1,i+1} \mapsto h_i.$$

Then ι is an isomorphism of U-modules.

Let a, b, c be elements of U. Then the Jacobi identity $[a,\, [b,\, c]] + [b,\, [c,\, a]] + [c,\, [a,\, b]] = 0$ holds and thus $ad\left([a,\, b]\right) = [ad(a),\, ad(b)]$. In particular, the defining relations of U imply that $ad(e_i), ad(f_i)$ and $ad(h_i)$ all satisfy the same defining relations. Hence (1) is proved. The Jacobi identity also implies the following Leibniz rule.

$$ad(x)[a,\, b] = [ad(x)a,\, b] + [a,\, ad(x)b] \quad (x, a, b \in U)$$

In order to prove (2) and (3), we establish explicit formulas for $[x,\, L_{jk}]$ $(x = e_i, f_i, h_i)$ in the following steps.

(step 1) $[\, h_i,\, L_{jk}\,] = (\delta_{ij} - \delta_{i+1,j} - \delta_{ik} + \delta_{i+1,k})L_{jk} \quad (j \neq k).$

EXERCISE 2.3. Prove this formula.

(step 2) $[e_i,\, L_{ik}] = 0 \quad (k \neq i).$

For $k = i \pm 1, i+2$, this is a direct consequence of the defining relations. For other values of k, it is proved by induction.

(step 3) $[\, e_i,\, L_{j,i+1}\,] = 0 \quad (j \neq i+1).$

The proof is the same as (step 2).

(step 4) *The following hold. In particular, we have* $[e_i,\, L_{jk}] \in L$ *for any* j, k.

(i) $[e_i,\, L_{jk}] = 0 \quad (j \neq i+1,\, k \neq i, j \neq k).$
(ii) $[e_i,\, L_{i+1,k}] = L_{ik} \quad (k \neq i, i+1).$
(iii) $[e_i,\, L_{ji}] = -L_{j,i+1} \quad (j \neq i+1, i).$
(iv) $[e_i,\, L_{i+1,i}] = h_i.$

We start with (i). It is obvious for the cases $j > k$, $i < j < k$ and $j < k < i$. The cases $j = i < k$ and $j < i = k-1 < k$ are nothing but (step 2) and (step 3). In the remaining case $j < i < k-1$, we know that $[e_i,\, L_{jk}] = 0$ by the following computation.

$$\begin{aligned}
[e_i,\, [L_{j,i+1},\, L_{i+1,k}]] &= [L_{j,i+1},\, [e_i,\, L_{i+1,k}]] = [L_{j,i+1},\, L_{ik}] \\
&= [[L_{ji},\, e_i],\, L_{ik}] = [[L_{ji},\, L_{ik}],\, e_i] \\
&= -[e_i,\, L_{jk}]
\end{aligned}$$

Next we prove (ii). It is obvious for the case $k \geq i+2$. If $k \leq i-1$, it follows from

$$[e_i,\, [f_i,\, L_{ik}]] = [h_i,\, L_{ik}] = L_{ik}.$$

The proof of (iii) is the same as (ii), and (iv) is obvious. Hence we have proved the assertion.

EXERCISE 2.4. Find formulas for $[f_i,\, L_{jk}]$ and prove them by using the automorphism of U defined by $e_i \mapsto f_i, h_i \mapsto -h_i, f_i \mapsto e_i$.

(step 5) *The space L is stable under $ad(h_i)$ and $ad(L_{ij})$. In other words, L is a Lie algebra.*

By (step 1), L is stable under $ad(h_i)$. We shall show by induction on $|j-i|$ that it is stable under $ad(L_{ij})$. If $i<j$ for example, we use (step 4) to conclude

$$[L_{ij},\, L] = [[e_i,\, L_{i+1,j}]\,,\, L] = [e_i,\, [L_{i+1,j},\, L]] + [L_{i+1,j},\, [e_i,\, L]] \subset L.$$

The argument for the case $i>j$ is similar.

(step 6) *We have an isomorphism of U-modules, $\iota : \mathfrak{g} \simeq L$.*

In fact, the above formulas imply that ι is a surjective homomorphism of U-modules. Since $\mathfrak{g}$ is an irreducible U-module by Assertion 2, it is enough to show that $L \neq 0$. But if $L=0$, we are forced to have $e_i = 0$, $f_i = 0$, $h_i = 0$ for all i, which contradicts the existence of the non-zero irreducible module of Assertion 2.

Assertion 4 *We have $[\iota(X),\, \iota(Y)] = \iota\left([X,\, Y]\right)$ $(X, Y \in \mathfrak{g})$.*

If one of X, Y is a diagonal matrix, the assertion follows from $[h_i,\, h_j] = 0$ and (step 1). To treat the other cases, we consider the root space decomposition $\mathfrak{g} = \mathfrak{h} \oplus (\oplus_{i\neq j} \mathbb{C} E_{ij})$ and the root system $\Phi = \{x_i - x_j\}_{i\neq j}$. Using the simple roots $\alpha_i = x_i - x_{i+1}$, we can describe Φ as follows.

$$\Phi = \left\{\pm(\alpha_i + \cdots + \alpha_{j-1})\right\}_{i<j}$$

For the root $\alpha = \alpha_i + \cdots + \alpha_{j-1}$ (resp. $\alpha = -(\alpha_i + \cdots + \alpha_{j-1})$), we denote its simultaneous eigenvector E_{ij} (resp. E_{ji}) by E_α.

For $\alpha, \beta \in \Phi$, we set

$$v_{\alpha,\beta} = [L_\alpha,\, L_\beta] - \iota\left([E_\alpha,\, E_\beta]\right).$$

By Assertion 3, we have $v_{\alpha,\beta} \in L$.

In the following, assume by way of contradiction that the set

$$\mathcal{A} = \{\, \gamma \in \bigoplus_{i=1}^{n-1} \mathbb{Z}\alpha_i \mid \gamma = \alpha + \beta,\ v_{\alpha,\beta} \neq 0\ (\exists\, \alpha, \beta \in \Phi)\}$$

is non-empty. Introduce a partial order on $\mathcal{A}$ as follows.

$$\gamma_1 \trianglerighteq \gamma_2 \Leftrightarrow \gamma_1 - \gamma_2 \in \bigoplus_{i=1}^{n-1} \mathbb{Z}_{\geq 0}\, \alpha_i$$

We take a maximal element γ with respect to the order and suppose that $\gamma = \alpha + \beta$ and $v_{\alpha,\beta} \neq 0$.

By Assertion 3(3), ι is an isomorphism of U-modules. Thus

$$\begin{aligned} e_i v_{\alpha,\beta} =& \big(\, [[L_{\alpha_i},\, L_\alpha]\,,\, L_\beta] - \iota\big([[E_{\alpha_i},\, E_\alpha]\,,\, E_\beta]\big)\big) \\ &+ \big(\, [\,L_\alpha,\, [\,L_{\alpha_i},\, L_\beta\,]\,] - \iota\big([\,E_\alpha,\, [\,E_{\alpha_i},\, E_\beta\,]\,]\big)\big). \end{aligned}$$

Note that if $\alpha = -\alpha_i$ and $\beta = \alpha_i$, then $v_{\alpha,\beta} = 0$, which is a contradiction. If $\alpha = -\alpha_i$ and $\beta \neq \alpha_i$, then we again have $v_{\alpha,\beta} = 0$ by the following computation for $j \neq i$ or $k \neq i+1$.

$$[f_i,\, L_{jk}] - \iota([E_{i+1,i},\, E_{jk}]) = [f_i,\, L_{jk}] - \delta_{ij} L_{i+1,k} + \delta_{i+1,k} L_{ji} = 0.$$

We similarly have $v_{\alpha,\beta} = 0$ for the case $\beta = -\alpha_i$. Hence both $\alpha = -\alpha_i$ and $\beta = -\alpha_i$ are impossible.

We divide the remaining cases into four subcases according to whether $\alpha+\alpha_i$ and $\beta+\alpha_i$ are elements of Φ or not, and show that $e_i v_{\alpha,\beta}=0$. We first consider the case where one of $\alpha+\alpha_i$, $\beta+\alpha_i$ belongs to Φ. As the argument is the same, we only treat the case where $\alpha+\alpha_i\in\Phi$. When $\alpha+\alpha_i\in\Phi$, we can write $[E_{\alpha_i}, E_\alpha]=c_{\alpha_i,\alpha}E_{\alpha_i+\alpha}$ with a non-zero number $c_{\alpha_i,\alpha}$. Thus,

$$\begin{aligned}[L_{\alpha_i}, L_\alpha] &= e_i\iota(E_\alpha)=\iota(e_iE_\alpha)=\iota([E_{\alpha_i}, E_\alpha])\\ &= c_{\alpha_i,\alpha}\iota(E_{\alpha_i+\alpha})=c_{\alpha_i,\alpha}L_{\alpha_i+\alpha}.\end{aligned}$$

This means that the element

$$[[L_{\alpha_i}, L_\alpha], L_\beta]-\iota\big([[E_{\alpha_i}, E_\alpha], E_\beta]\big)$$

is equal to a scalar multiple of $v_{\alpha+\alpha_i,\beta}$. By the maximality of γ, this element is 0. Using this we have

$$e_iv_{\alpha,\beta}=[L_\alpha, [L_{\alpha_i}, L_\beta]]-\iota\big([E_\alpha, [E_{\alpha_i}, E_\beta]]\big).$$

If $\beta+\alpha_i\notin\Phi$, then $\beta+\alpha_i\neq 0$ implies that $[E_{\alpha_i}, E_\beta]=0$. So

$$[L_{\alpha_i}, L_\beta]=e_i\iota(E_\beta)=\iota(e_iE_\beta)=\iota([E_{\alpha_i}, E_\beta])=0,$$

which implies that $e_iv_{\alpha,\beta}=0$. If $\beta+\alpha_i\in\Phi$, then $e_iv_{\alpha,\beta}$ is equal to a scalar multiple of $v_{\alpha,\beta+\alpha_i}$ and the maximality of γ again implies that $e_iv_{\alpha,\beta}=0$.

If both $\alpha+\alpha_i,\beta+\alpha_i$ are not in Φ, then

$$\begin{aligned}[L_{\alpha_i}, L_\alpha] &= e_i\iota(E_\alpha)=\iota(e_iE_\alpha)=\iota([E_{\alpha_i}, E_\alpha])=0,\\ [L_{\alpha_i}, L_\beta] &= e_i\iota(E_\beta)=\iota(e_iE_\beta)=\iota([E_{\alpha_i}, E_\beta])=0.\end{aligned}$$

Hence we have $e_iv_{\alpha,\beta}=0$ in this case also.

Since i is arbitrary and $e_iv_{\alpha,\beta}=0$, Assertion 3(3) tells us that $v_{\alpha,\beta}$ is equal to a scalar multiple of $\iota(E_{1n})$. In particular, we have

$$\gamma=\alpha_1+\cdots+\alpha_{n-1}$$

and $\{\alpha,\beta\}=\{\alpha_1+\cdots+\alpha_k,\alpha_{k+1}+\cdots+\alpha_{n-1}\}$ for some k. However, $[L_\alpha, L_\beta]$ is equal to L_{1n} in this case and so $v_{\alpha,\beta}=0$, which contradicts our choice of γ. Hence, we have $\mathcal{A}=\emptyset$ and Assertion 4 follows.

We are now ready to prove Theorem 2.2. We have constructed a map $\iota:\mathfrak{g}\to U$ which satisfies $[\iota(X), \iota(Y)]=\iota([X, Y])$. Our next task is to prove the universality of the pair (U,ι); however, this is obvious because the map $\phi: U\to A$ is uniquely determined by the requirements that $\phi(e_i)=\rho(E_{i,i+1})$ etc. □

2.2. The quantum algebra of type A_{r-1}

Based on Theorem 2.2 Drinfeld and Jimbo introduced the quantum algebra which is obtained as a "deformation"of the enveloping algebra of $\mathfrak{sl}_r=\mathfrak{sl}(r,\mathbb{C})$. The definition is as follows. We choose $\mathbb{Q}(v)$ as a base field since it is not necessary to assume it to be $\mathbb{C}(v)$. The element t_i is often denoted by v^{h_i} and $\alpha_j(h_i)=2\delta_{ij}-\delta_{i,j+1}-\delta_{i+1,j}$ by definition.

DEFINITION 2.5. Let $K=\mathbb{Q}(v)$ where v is an indeterminate. The **quantum algebra of type** A_{r-1} is the unital associative K-algebra $U_v(\mathfrak{sl}_r)$ defined by the following generators and relations.

Generators: $t_i^{\pm1}, e_i, f_i \quad (1\le i\le r-1)$.

Relations:

$$t_i e_j t_i^{-1} = v^{\alpha_j(h_i)} e_j, \quad t_i f_j t_i^{-1} = v^{-\alpha_j(h_i)} f_j,$$

$$[e_i, f_j] = \delta_{ij} \frac{t_i - t_i^{-1}}{v - v^{-1}},$$

$$[t_i, t_j] = 0, \quad t_i t_i^{-1} = t_i^{-1} t_i = 1,$$

$$e_i^2 e_j - (v + v^{-1}) e_i e_j e_i + e_j e_i^2 = 0 \ (i - j = \pm 1),$$
$$e_i e_j = e_j e_i \ (otherwise),$$

$$f_i^2 f_j - (v + v^{-1}) f_i f_j f_i + f_j f_i^2 = 0 \ (i - j = \pm 1),$$
$$f_i f_j = f_j f_i \ (otherwise).$$

These relations are called the (deformed) Serre relations.

DEFINITION 2.6. Let $[k] = \frac{v^k - v^{-k}}{v - v^{-1}}$, for $k \in \mathbb{N}$, and let $[n]! = \prod_{k=1}^{n} [k]$. Then $f_i^{(n)}$ is defined by

$$f_i^{(n)} = \frac{f_i^n}{[n]!}.$$

Roughly speaking, the quantum algebra is the algebra which is obtained by "integrating"the Cartan subalgebra and deforming the other relations "nicely".

We may obtain representations of $U_v(\mathfrak{sl}_r)$ by deforming the representations of $\mathfrak{g}$. We can also define tensor product representations by deforming the coproduct of the enveloping algebra as follows.

$$\Delta(t_i) = t_i \otimes t_i, \ \Delta(e_i) = 1 \otimes e_i + e_i \otimes t_i^{-1},$$
$$\Delta(f_i) = f_i \otimes 1 + t_i \otimes f_i.$$

EXERCISE 2.7. Verify that Δ defines an algebra homomorphism from $U_v(\mathfrak{sl}_r)$ to $U_v(\mathfrak{sl}_r) \otimes U_v(\mathfrak{sl}_r)$.

EXERCISE 2.8. Let $V = K^r$ and define $\rho : U_v(\mathfrak{sl}_r) \to \mathrm{End}(V)$ by

$$\rho(t_i) = \mathrm{I} + (v - 1) E_{i,i} + (v^{-1} - 1) E_{i+1,i+1},$$
$$\rho(e_i) = E_{i,i+1}, \quad \rho(f_i) = E_{i+1,i}.$$

Show that (ρ, V) is a representation of $U_v(\mathfrak{sl}_r)$. This is called the natural (or vector) representation of $U_v(\mathfrak{sl}_r)$.

EXERCISE 2.9. Let V be the natural representation of $U_v(\mathfrak{sl}_2)$. Decompose $V \otimes V$ into a sum of irreducible $U_v(\mathfrak{sl}_2)$-submodules.

CHAPTER 3

Kac-Moody Lie algebras

3.1. Lie algebras by generators and relations

In the last chapter we have exhibited the enveloping algebra of the special linear Lie algebra $\mathfrak{sl}(n, \mathbb{C})$ as an algebra defined by generators and relations. For the rest of these lectures we consider the quantum algebra which is obtained as a deformation of the enveloping algebra of the following Lie algebra:

$$\mathfrak{g} = \mathfrak{sl}(r, \mathbb{C}) \otimes \mathbb{C}[t, t^{-1}] \oplus \mathbb{C}c \oplus \mathbb{C}d,$$

where the Lie bracket is given by

$$[X \otimes t^n, Y \otimes t^m] = [X, Y] \otimes t^{n+m} + n(trXY)\delta_{0,n+m}c,$$

$$[c, X \otimes t^n] = 0, \quad [d, X \otimes t^n] = nX \otimes t^n, \quad [c, d] = 0.$$

EXERCISE 3.1. Check that $\mathfrak{g}$ is a Lie algebra.

The purpose of this chapter is to give another definition of this Lie algebra by generators and relations. This definition leads to a definition of the enveloping algebra of $\mathfrak{g}$ by generators and relations. We do this in order to introduce the deformation $U_v(\mathfrak{g})$. We can carry out this program since this Lie algebra belongs to a class of Lie algebras introduced independently by Kac and Moody. They started with a definition which is very close to the definition by generators and relations. Then using this definition they clarified the structure of the Lie algebras and their representations. The desired defining relations were obtained from this process. In this chapter, we learn a little bit about the theory of Kac-Moody Lie algebras. We first consider some general concepts.

DEFINITION 3.2. For a vector space V, we view its tensor algebra $T(V)$ as a Lie algebra with bracket $[a, b] = ab - ba$. The **free Lie algebra** $\mathfrak{g}(V)$ generated by V is the minimal Lie subalgebra of $T(V)$ which contains V.

It is obvious that the Lie algebra $\mathfrak{g}(V)$ exists because $\mathfrak{g}(V)$ is the intersection of all Lie subalgebras of $T(V)$ which contain V.

LEMMA 3.3. *We have* $\mathfrak{g}(V) = \sum_{N \geq 0} (ad\ V)^N V$.

PROOF. If $\mathfrak{A} \subset T(V)$ is a Lie subalgebra which contains V, then

$$\mathfrak{A} \supset \sum_{N \geq 0} (ad\ V)^N V.$$

Hence, $\mathfrak{g}(V) \supset \sum_{N \geq 0} (ad\ V)^N V$. To see that equality holds it is enough to observe that $\sum_{N \geq 0} (ad\ V)^N V$ is itself a Lie subalgebra. We prove by induction on i that

$$[(ad\ V)^i V, (ad\ V)^j V] \subset \sum_{N \geq 0} (ad\ V)^N V.$$

When $i = 0$ there is nothing to prove. Assume that the claim is true for i. Then $[(ad\, V)^{i+1}V, (ad\, V)^j V]$ is contained in

$$ad\, V([(ad\, V)^i V, (ad\, V)^j V]) + [(ad\, V)^i V, (ad\, V)^{j+1} V]$$

and the claim follows by induction. □

DEFINITION 3.4. Let R be a subspace of $\mathfrak{g}(V)$. We denote by $\mathcal{I}(R)$ the minimal ideal of the Lie algebra $\mathfrak{g}(V)$ which contains R. Then the quotient space $\mathfrak{g} = \mathfrak{g}(V)/\mathcal{I}(R)$ is a Lie algebra. This Lie algebra is the **Lie algebra defined by generators V and relations R.**

If $R = \sum \mathbb{C}R_i$ $(R_i \in \mathfrak{g}(V))$ then we usually write $R_1 = 0, R_2 = 0, \dots$ instead of giving the space R.

The proof of the following lemma is the same as Lemma 3.3.

LEMMA 3.5. *We have* $\mathcal{I}(R) = \sum_{N \geq 0} (ad\, V)^N R$.

The relation between a Lie algebra defined by generators and relations and its enveloping algebra is given by the following proposition.

PROPOSITION 3.6. *Let $\mathfrak{g}$ be the Lie algebra defined by generators V and relations R and let I be the minimal two-sided ideal of $T(V)$ which contains R. Then there is an isomorphism of algebras $U(\mathfrak{g}) \simeq T(V)/I$. In particular, $I \cap \mathfrak{g}(V) = \mathcal{I}(R)$.*

PROOF. Let $\iota : \mathfrak{g} \to T(V)/I$ be the linear map defined by

$$\iota(X \pmod{\mathcal{I}(R)}) = X \pmod{I} \quad (X \in \mathfrak{g}(V)).$$

By Lemma 3.5, we have $\mathcal{I}(R) \subset I$ so this map is well-defined. In addition, $\iota([X, Y]) = [\iota(X), \iota(Y)]$. We show the universality of the pair $(T(V)/I, \iota)$.

Let $\rho : \mathfrak{g} \to A$ be a linear map which satisfies $\rho([X, Y]) = [\rho(X), \rho(Y)]$. By setting $\phi(X) = \rho(X \pmod{\mathcal{I}(R)})$ for $X \in V$, we have an algebra homomorphism $\phi : T(V) \to A$. Using the property $\rho([X, Y]) = [\rho(X), \rho(Y)]$, if $v_1, \dots, v_N, v_{N+1} \in V$ then

$$\begin{aligned}
\phi(ad(v_1) \cdots ad(v_N) v_{N+1}) &= ad(\phi(v_1)) \cdots ad(\phi(v_N)) \phi(v_{N+1}) \\
&= [\rho(v_1 \pmod{\mathcal{I}(R)}), [\cdots, [\rho(v_N \pmod{\mathcal{I}(R)}), \rho(v_{N+1} \pmod{\mathcal{I}(R)})] \cdots]] \\
&= \rho(ad(v_1) \cdots ad(v_N) v_{N+1} \pmod{\mathcal{I}(R)}).
\end{aligned}$$

Taking Lemma 3.3 into consideration, this implies that if $X \in \mathfrak{g}(V)$ then $\phi(X) = \rho(X \pmod{\mathcal{I}(R)})$.

Recalling that $R \subset \mathcal{I}(R) \subset \mathfrak{g}(V)$ we have $\phi(R) = 0$. Since ϕ is an algebra homomorphism, we have obtained $\phi(I) = 0$. Therefore, there is a well-defined algebra homomorphism $\overline{\phi} : T(V)/I \to A$ which satisfies

$$\overline{\phi} \circ \iota(X) = \phi(X \pmod{I}) = \rho(X) \quad (X \in \mathfrak{g}).$$

The uniqueness of $\overline{\phi}$ follows from the fact that the values of $\phi(V)$ are determined by ρ. Hence $(T(V)/I, \iota)$ is universal, and we have proved that $U(\mathfrak{g}) \simeq T(V)/I$. □

3.2. Kac-Moody Lie algebras

To define Kac-Moody Lie algebras, we need the notion of the (generalized) Cartan matrix and its realization. We start with these definitions.

DEFINITION 3.7. A square matrix with integer entries $A = (a_{ij})$ is called a **Cartan matrix** if it satisfies the three conditions: $a_{ii} = 2$, $a_{ij} \leq 0$ $(i \neq j)$ and $a_{ij} = 0 \Leftrightarrow a_{ji} = 0$.

DEFINITION 3.8. A **realization** of a Cartan matrix $A = (a_{ij})_{0 \leq i,j < r}$ is a triple $(\mathfrak{h}, \Pi, \Pi^\vee)$ of a vector space $\mathfrak{h}$ and two sets of linearly independent vectors $\Pi = \{\alpha_i\}_{0 \leq i < r} \subset \mathfrak{h}$ and $\Pi^\vee = \{h_j\}_{0 \leq j < r} \subset \mathfrak{h}^*$ which satisfy

$$\dim \mathfrak{h} = 2r - \operatorname{rank} A, \quad \alpha_i(h_j) = a_{ji}.$$

In these lectures, we consider the following Cartan matrix, which is the **Cartan matrix of type $A^{(1)}_{r-1}$**.

$$A = \begin{pmatrix} 2 & -1 & & & -1 \\ -1 & 2 & -1 & & \\ \cdots & \cdots & \cdots & \cdots & \cdots \\ -1 & & & -1 & 2 \end{pmatrix} \quad (r \geq 3), \qquad \begin{pmatrix} 2 & -2 \\ -2 & 2 \end{pmatrix} \quad (r = 2).$$

Its realization for the case $r \geq 3$ is given by

$$\mathfrak{h} = \left(\bigoplus_{i=0}^{r-1} \mathbb{C} h_i \right) \oplus \mathbb{C} d,$$

$$\mathfrak{h}^* = \left(\bigoplus_{i=0}^{r-1} \mathbb{C} \alpha_i \right) \oplus \mathbb{C} \Lambda_0,$$

and $\Pi = \{\alpha_i\}_{0 \leq i \leq r-1}$, $\Pi^\vee = \{h_j\}_{0 \leq j \leq r-1}$. Here, α_i, Λ_0; h_j, d are linearly independent elements which form bases of $\mathfrak{h}$ and $\mathfrak{h}^*$ respectively, and satisfy

$$\alpha_i(h_j) = \begin{cases} 2 & (i = j) \\ -1 & (i - j \equiv \pm 1 \pmod{r}) \\ 0 & (\textit{otherwise}) \end{cases}$$

and

$$\alpha_i(d) = \delta_{i0}, \quad \Lambda_0(h_j) = \delta_{0j}, \quad \Lambda_0(d) = 0.$$

The case $r = 2$ is defined similarly. (See [**K**, 1.1] for the theory of Cartan matrices and their realizations.)

Kac and Moody have defined the Lie algebra associated with a Cartan matrix as follows. In our case this Lie algebra is called the **Kac-Moody Lie algebra** of type $A^{(1)}_{r-1}$. The existence of $\mathcal{R}_{\max}$ in the definition will be verified in Proposition 3.12.

DEFINITION 3.9. For a realization $(\mathfrak{h}, \Pi, \Pi^\vee)$ of a Cartan matrix A, we let $\tilde{\mathfrak{g}}(V)$ be the Lie algebra defined by

Generators: $$\left(\bigoplus_{i=0}^{r-1} \mathbb{C} e_i \right) \oplus \mathfrak{h} \oplus \left(\bigoplus_{i=0}^{r-1} \mathbb{C} f_i \right).$$

Relations:

$$[h, e_i] = \alpha_i(h) e_i, \quad [h, f_i] = -\alpha_i(h) f_i,$$
$$[e_i, f_j] = \delta_{ij} h_i, \quad [h, h'] = 0 \ (h, h' \in \mathfrak{h}).$$

Let $\mathcal{R}_{\max}$ be the maximal ideal of $\tilde{\mathfrak{g}}(V)$ among those ideals $\mathcal{R}$ which satisfy $\mathcal{R}\cap\mathfrak{h} = 0$. We then call $\tilde{\mathfrak{g}}(V)/\mathcal{R}_{\max}$ the **Kac-Moody Lie algebra** associated with the Cartan matrix A, and denote it by $\mathfrak{g}(A)$. The simultaneous eigenspace decomposition of $\mathfrak{g}(A)$ with respect to $ad(\mathfrak{h})$ is called the **root space decomposition** and the non-zero simultaneous eigenvalues are called **roots**.

The Kac-Moody Lie algebra is not defined by generators and relations, but we have the following theorem; see [**K**, Theorem 9.11].

THEOREM 3.10. *We assume that A is the Cartan matrix of type $A^{(1)}_{r-1}$ with $r \geq 3$ for simplicity. Then the Lie algebra $\mathfrak{g}(A)$ is defined by the following generators and relations.*

Generators:

$$\left(\bigoplus_{i=0}^{r-1}\mathbb{C}e_i\right)\oplus\mathfrak{h}\oplus\left(\bigoplus_{i=0}^{r-1}\mathbb{C}f_i\right).$$

Relations:

$$[h,\, e_i] = \alpha_i(h)e_i, \quad [h,\, f_i] = -\alpha_i(h)f_i,$$

$$[e_i,\, f_j] = \delta_{ij}h_i, \quad [h,\, h'] = 0\ (h, h' \in \mathfrak{h}),$$

$$[e_i,\, [e_i,\, e_j]] = 0,\ [f_i,\, [f_i,\, f_j]] = 0\ \ (i - j \equiv \pm 1 \pmod r),$$
$$[e_i,\, e_j] = 0, \quad [f_i,\, f_j] = 0\ \ (\textit{otherwise}).$$

We do not explain its proof in detail, but give a rough sketch of the argument to convince the reader.

PROPOSITION 3.11. *The relations stated in Theorem 3.10 all hold in $\mathfrak{g}(A)$.*

PROOF. For example, if i, j satisfy $i - j \equiv \pm 1 \pmod r$ we define $\mathcal{R}$ as follows.

$$\mathcal{R} = \sum_{N\geq 0}(ad\,(\bigoplus_{i=0}^{r-1}\mathbb{C}f_i))^N(f_i^2f_j - 2f_if_jf_i + f_jf_i^2)$$

Since $[e_k,\, f_i^2f_j - 2f_if_jf_i + f_jf_i^2] = 0\ (0 \leq k \leq r-1)$, $\mathcal{R}$ is an ideal and $\mathcal{R}\cap\mathfrak{h} = 0$. Thus we have $\mathcal{R} \subset \mathcal{R}_{\max}$, and the relation

$$f_i^2f_j - 2f_if_jf_i + f_jf_i^2 = 0$$

holds in $\mathfrak{g}(A)$. The other relations are proved in a similar way. □

This proposition says that $\mathfrak{g}(A)$ is a quotient Lie algebra of the Lie algebra defined by the generators and the relations given in Theorem 3.10. We denote this Lie algebra by $\hat{\mathfrak{g}}(A)$, and assume that $\mathfrak{g}(A)$ is the quotient of $\hat{\mathfrak{g}}(A)$ by an ideal $\mathfrak{R}$. Our goal is to prove that $\mathfrak{R} = 0$.

We assume to the contrary that $\mathfrak{R}$ is not 0. To show that this leads to a contradiction we apply $e^{ad(f_i)}e^{-ad(e_i)}e^{ad(f_i)}$ to a simultaneous eigenvector of $ad(\mathfrak{h})$. Since $ad(e_i)$ and $ad(f_i)$ are locally nilpotent, the expression is well-defined. The result is again a simultaneous eigenvector and the simultaneous eigenvalue changes from $\alpha \in \mathfrak{h}^*$ to $s_i\alpha := \alpha - \alpha(h_i)\alpha_i$. In particular, if we denote by $\Phi_{\mathfrak{R}}$ the set of roots appearing in $\mathfrak{R}$, we have $s_i\Phi_{\mathfrak{R}} = \Phi_{\mathfrak{R}}$. Using this fact, we can prove that a root $\alpha = \sum n_i\alpha_i \in \Phi_{\mathfrak{R}}$ attains the minimal height ($\mathrm{ht}(\alpha) = |\sum n_i|$) of $\Phi_{\mathfrak{R}}$ only when $\alpha = \pm\alpha_i$ for some i or $\alpha(h_i) \leq 0$ for all i. The former cannot happen since

e_i, f_i are not in $\mathfrak{R}$, and the latter occurs only when $\alpha = n\delta$ ($\delta = \sum_{i=0}^{r-1}\alpha_i$). Hence we have $n\delta \in \Phi_{\mathcal{R}}$ for some n.

The Lie algebra $\tilde{\mathfrak{g}}(V)$ has a triangular decomposition, which will be explained in Proposition 3.12, and its roots are in $\pm(\oplus\mathbb{Z}_{\geq 0}\alpha_i)$. Hence we have a decomposition of $\mathfrak{R}$ into the spaces with positive roots and negative roots as follows.

$$\mathfrak{R} = \mathfrak{R}_+ \oplus \mathfrak{R}_-$$

If $\mathfrak{R}_- \neq 0$ we can prove that there exists an embedding of $\mathfrak{R}_-/[\mathfrak{R}_-, \mathfrak{R}_-]$ into a certain direct sum of modules called Verma modules. We apply "the Casimir operator"to conclude that an element of $\Phi_{\mathfrak{R}_-}$ cannot have the form $n\delta$. The same argument works if $\mathfrak{R}_+ \neq 0$, and we have reached a contradiction.

The readers who are interested in the details of the proof should consult [**K**, Proposition 9.11]. The proof is long and it may be wise to proceed without stopping at this point.

We now explain why the Kac-Moody Lie algebras are well-defined and describe their triangular decomposition. The triangular decomposition plays an important role in the treatment of "highest weight modules"in later sections. The enveloping algebra also has a triangular decomposition and we will meet an analogue of the triangular decomposition when we treat the quantum algebra.

PROPOSITION 3.12 ([**K**, Thorem 1.2]). *The following hold.*

(1) *Let A be a Cartan matrix, $\mathfrak{g}(A)$ the Kac-Moody Lie algebra associated with A. We set $V_+ = \oplus_{i=0}^{r-1}\mathbb{C}e_i$, $V_- = \oplus_{i=0}^{r-1}\mathbb{C}f_i$ and let $\mathfrak{g}(V)$ be the free Lie algebra generated by $V = V_+ \oplus \mathfrak{h} \oplus V_-$. We define its quotient Lie algebra $\tilde{\mathfrak{g}}(V)$ as in Definition 3.9. Then we have the following direct sum decomposition, called the triangular decomposition of $\tilde{\mathfrak{g}}(V)$.*

$$\tilde{\mathfrak{g}}(V) = \mathfrak{g}(V_+) \oplus \mathfrak{h} \oplus \mathfrak{g}(V_-)$$

Each direct summand has a simultaneous eigenspace decomposition with respect to ad($\mathfrak{h}$). The set of the simultaneous eigenvalues is $(\oplus\mathbb{Z}_{\geq 0}\alpha_i)\setminus\{0\}$ in $\mathfrak{g}(V_+)$, $\{0\}$ in $\mathfrak{h}$ and $(\oplus\mathbb{Z}_{\leq 0}\alpha_i)\setminus\{0\}$ in $\mathfrak{g}(V_-)$.

(2) *There exists a unique maximal ideal $\mathcal{R}_{\max}$ of $\tilde{\mathfrak{g}}(V)$ among those ideals whose intersection with $\mathfrak{h}$ is 0.*

(3) *Set $\mathcal{R}_\pm = \mathcal{R}_{\max} \cap \mathfrak{g}(V_\pm)$. Then we have the* **triangular decomposition**

$$\mathfrak{g}(A) = (\mathfrak{g}(V_+)/\mathcal{R}_+) \oplus \mathfrak{h} \oplus (\mathfrak{g}(V_-)/\mathcal{R}_-).$$

(4) *Consider the simultaneous eigenspace decomposition (root space decomposition) of $\mathfrak{g}(A)$ with respect to ad($\mathfrak{h}$). Then a simultaneous eigenvector has eigenvalue zero if and only if it belongs to $\mathfrak{h}$. Let Φ be the set of roots and set $\Phi_+ = \Phi \cap (\oplus\mathbb{Z}_{\geq 0}\alpha_i)$ and $\Phi_- = \Phi \cap (\oplus\mathbb{Z}_{\leq 0}\alpha_i)$. Then $\Phi = \Phi_+ \sqcup \Phi_-$.*

PROOF. (1) This is a consequence of the following three assertions.

Assertion 1 *Set $\tilde{\mathfrak{n}}_\pm = \sum_{N\geq 0}(ad\ V_\pm)^N V_\pm$. Then*

$$\tilde{\mathfrak{g}}(V) = \tilde{\mathfrak{n}}_+ + \mathfrak{h} + \tilde{\mathfrak{n}}_-.$$

Because of inclusions $[V, \tilde{\mathfrak{n}}_+ + \mathfrak{h} + \tilde{\mathfrak{n}}_-] \subset \tilde{\mathfrak{n}}_+ + \mathfrak{h} + \tilde{\mathfrak{n}}_-$ in $\tilde{\mathfrak{g}}(V)$ and $V \subset \tilde{\mathfrak{n}}_+ + \mathfrak{h} + \tilde{\mathfrak{n}}_-$, we have $\tilde{\mathfrak{g}}(V) \subset \tilde{\mathfrak{n}}_+ + \mathfrak{h} + \tilde{\mathfrak{n}}_-$ by Lemma 3.3. The opposite inclusion is obvious.

Assertion 2 *We have $\mathfrak{g}(V_\pm) \simeq \tilde{\mathfrak{n}}_\pm$. In other words, $\tilde{\mathfrak{n}}_\pm$ is the free Lie algebra generated by $V_\pm$.*

We only show that $\mathfrak{g}(V_-) \simeq \tilde{\mathfrak{n}}_-$. Let $\lambda \in \mathfrak{h}^*$ and let $\tilde{M}(\lambda)$ be the tensor algebra $T(\oplus_{i=0}^{r-1}\mathbb{C}\xi_i)$ of an r-dimensional vector space $\oplus_{i=0}^{r-1}\mathbb{C}\xi_i$. Then we define operators e_i, f_i, for $0 \leq i \leq r-1$, and $h \in \mathfrak{h}$ on $\tilde{M}(\lambda)$ as follows.

$$f_i \cdot a = \xi_i \otimes a, \quad h \cdot 1 = \lambda(h)1, \quad e_i \cdot 1 = 0,$$
$$h \cdot (\xi_i \otimes a) = -\alpha_i(h)\xi_i \otimes a + \xi_i \otimes (h \cdot a),$$
$$e_i \cdot (\xi_j \otimes a) = \delta_{ij} h_i \cdot a + \xi_j \otimes (e_i \cdot a).$$

By checking the defining relations, we know that $\tilde{M}(\lambda)$ is a $\tilde{\mathfrak{g}}(V)$-module. We consider the composite of $\tilde{\mathfrak{n}}_- \to \tilde{M}(\lambda) : n \mapsto n \cdot 1$ with the natural surjection $\mathfrak{g}(V_-) \to \tilde{\mathfrak{n}}_-$. Since this map $F(f_0, \ldots, f_{r-1}) \mapsto F(\xi_0, \ldots, \xi_{r-1})$ is injective, $\mathfrak{g}(V_-) \to \tilde{\mathfrak{n}}_-$ is also, and the assertion follows. Note that we have also proved that the map $\tilde{\mathfrak{n}}_- \to \tilde{M}(\lambda)$ is injective.

Assertion 3 *We have $\tilde{\mathfrak{g}}(V) = \mathfrak{g}(V_+) \oplus \mathfrak{h} \oplus \mathfrak{g}(V_-)$.*

By virtue of the previous two assertions, it is enough to observe that this sum is direct. Assume that the sum of $n_- \in \mathfrak{g}(V_-)$, $h \in \mathfrak{h}$ and $n_+ \in \mathfrak{g}(V_+)$ is 0. Then we have

$$(n_- \cdot 1) + (h \cdot 1) + (n_+ \cdot 1) = (n_- \cdot 1) + \lambda(h)1 = 0$$

for all λ, from which we have $h = 0$. Since we have proved in Assertion 2 that $n_- = 0$ whenever $n_- \cdot 1 = 0$, this proves (1).
(2) Since the Lie algebra $\tilde{\mathfrak{g}}(V)$ has $ad(\mathfrak{h})$- simultaneous eigenvectors as a basis, any ideal has a simultaneous eigenspace decomposition $\mathcal{R} = \oplus_\alpha \mathcal{R}_\alpha$, too. Hence the sum $\sum_{\mathcal{R}\cap\mathfrak{h}=0} \mathcal{R}$ of all ideals $\mathcal{R}$ which satisfy $\mathcal{R} \cap \mathfrak{h} = 0$ also has the property that its intersection with $\mathfrak{h}$ is 0. This ideal is $\mathcal{R}_{\max}$ and the assertion follows.

Parts (3) and (4) follow from (1) and (2). □

We now show that the Lie algebra $\mathfrak{sl}(r, \mathbb{C}) \otimes \mathbb{C}[t, t^{-1}] \oplus \mathbb{C}c \oplus \mathbb{C}d$ is the Kac-Moody Lie algebra of type $A_{r-1}^{(1)}$.

PROPOSITION 3.13 ([**K**, Proposition 1.4(a)]). *Let $\mathfrak{g}$ be a Lie algebra and assume that there exist (i) a commutative Lie subalgebra $\mathfrak{h}$, (ii) a set $\{e_i, f_i\}_{0 \leq i < r}$ of elements in $\mathfrak{g}$, (iii) a set $\Pi = \{\alpha_i\}_{0 \leq i < r} \subset \mathfrak{h}^*$, (iv) a set $\Pi^\vee = \{h_j\}_{0 \leq i < r} \subset \mathfrak{h}$, which satsify the following conditions.*

1. *The Lie algebra $\mathfrak{g}$ is generated by $\{e_0, f_0, \ldots, e_{r-1}, f_{r-1}, \mathfrak{h}\}$.*
2. *The equations $[e_i, f_j] = \delta_{ij} h_i$, $[h, e_i] = \alpha_i(h) e_i$, $[h, f_i] = -\alpha_i(h) f_i$ hold.*
3. *The triple $(\mathfrak{h}, \Pi, \Pi^\vee)$ is a realization of a Cartan matrix A.*
4. *If an ideal $\mathcal{R}$ of $\mathfrak{g}$ satisfies $\mathcal{R} \cap \mathfrak{h} = 0$, then $\mathcal{R} = 0$.*

Then $\mathfrak{g}$ is isomorphic to the Kac-Moody Lie algebra associated with the Cartan matrix A.

PROOF. Set $V = \left(\oplus_{i=1}^{r-1}\mathbb{C}e_i\right) \oplus \mathfrak{h} \oplus \left(\oplus_{i=1}^{r-1}\mathbb{C}f_i\right)$. Then conditions (1) and (2) define a surjective Lie algebra homomorphism $\phi : \tilde{\mathfrak{g}}(V) \to \mathfrak{g}$ and, if we restrict ϕ to $\mathfrak{h}$, we know by the triangular decomposition of $\tilde{\mathfrak{g}}(V)$ that ϕ is an isomorphism. We now recall that the simultaneous eigenvalues of $\mathcal{R}_{\max}$ with respect to $ad(\mathfrak{h})$ are all non-zero. In particular, we have $\phi(\mathcal{R}_{\max}) \cap \mathfrak{h} = 0$. Now (4) implies that $\phi(\mathcal{R}_{\max}) = 0$. So, ϕ induces a surjective Lie algebra homomorphism $\mathfrak{g}(A) \to \mathfrak{g}$. Since its kernel is an ideal satisfying $\mathcal{R} \cap \mathfrak{h} = 0$, it must be 0. Hence, $\mathfrak{g}(A) \simeq \mathfrak{g}$. □

EXERCISE 3.14. In the Lie algebra $\mathfrak{g} = \mathfrak{sl}(r, \mathbb{C}) \otimes \mathbb{C}[t, t^{-1}] \oplus \mathbb{C}c \oplus \mathbb{C}d$, we choose e_i, f_i, h_i for $0 \le i \le r-1$ as follows.

$$e_0 = E_{n1} \otimes t, \quad e_i = E_{i,i+1} \otimes 1 \ (1 \le i \le r-1),$$

$$f_0 = E_{1n} \otimes t^{-1}, \quad f_i = E_{i+1,i} \otimes 1 \ (1 \le i \le r-1),$$

$$h_0 = -\sum_{i=1}^{r-1} h_i + c, \quad h_i = (E_{i,i} - E_{i+1,i+1}) \otimes 1 \ (i \ge 1).$$

Define $\alpha_i(h)$ suitably and show that $\mathfrak{g}$ satisfies the assumptions of Proposition 3.13.

By virtue of Exercise 3.14 and Proposition 3.13 we know that the Lie algebra $\mathfrak{sl}(r, \mathbb{C}) \otimes \mathbb{C}[t, t^{-1}] \oplus \mathbb{C}c \oplus \mathbb{C}d$ is the Kac-Moody Lie algebra of type $A_{r-1}^{(1)}$. By Theorem 3.10 we have also obtained its presentation in terms of generators and relations.

EXERCISE 3.15. Give another proof of Theorem 2.2 by showing that $\mathfrak{sl}(r, \mathbb{C})$ is a Kac-Moody Lie algebra.

3.3. The quantum algebra of type $A_{r-1}^{(1)}$

Based on the explanation given so far, in particular Proposition 3.6, the reader may feel it natural to define the algebra below as a deformation of the enveloping algebra of $\mathfrak{sl}(r, \mathbb{C}) \otimes \mathbb{C}[t, t^{-1}] \oplus \mathbb{C}c \oplus \mathbb{C}d$. In this definition, as in the case of type A_{r-1}, we choose $\mathbb{Q}(v)$ as the base field and the element t_i is often denoted by v^{h_i}.

DEFINITION 3.16. Let $K = \mathbb{Q}(v)$ and assume that $r \ge 3$. The **quantum algebra** U_v of type $A_{r-1}^{(1)}$ is the unital associative K-algebra defined by the following generators and relations.

Generators: $\quad t_i^{\pm 1}, v^{\pm d}, e_i, f_i \quad (0 \le i \le r-1).$

Relations:

$$t_i e_j t_i^{-1} = v^{\alpha_j(h_i)} e_j, \quad t_i f_j t_i^{-1} = v^{-\alpha_j(h_i)} f_j,$$

$$v^d e_j v^{-d} = v^{\delta_{0j}} e_j, \quad v^d f_j v^{-d} = v^{-\delta_{0j}} f_j,$$

$$[e_i, f_j] = \delta_{ij} \frac{t_i - t_i^{-1}}{v - v^{-1}},$$

$$v^d v^{-d} = v^{-d} v^d = 1, \quad t_i t_i^{-1} = t_i^{-1} t_i = 1,$$

$$[v^d, t_i] = 0, \quad [t_i, t_j] = 0,$$

$$e_i^2 e_j - (v + v^{-1}) e_i e_j e_i + e_j e_i^2 = 0 \ (i - j \equiv \pm 1 \ (\mathrm{mod}\ r)),$$

$$e_i e_j = e_j e_i \ (\textit{otherwise}),$$

$$f_i^2 f_j - (v + v^{-1}) f_i f_j f_i + f_j f_i^2 = 0 \ (i - j \equiv \pm 1 \ (\mathrm{mod}\ r)),$$

$$f_i f_j = f_j f_i \ (\textit{otherwise}).$$

Next assume that $r = 2$. Then the last four relations are replaced by

$$e_i^3 e_j - (v^2+1+v^{-2}) e_i^2 e_j e_i + (v^2+1+v^{-2}) e_i e_j e_i^2 - e_j e_i^3 = 0 \ (i \neq j),$$
$$f_i^3 f_j - (v^2+1+v^{-2}) f_i^2 f_j f_i + (v^2+1+v^{-2}) f_i f_j f_i^2 - f_j f_i^3 = 0 \ (i \neq j).$$

REMARK 3.17. When we consider the case $r = 1$, we understand that $U_v = U_v(\mathfrak{sl}_2)$. We define $f_i^{(n)}$ to be $\frac{f_i^n}{[n]!}$ as before.

CHAPTER 4

Crystal bases of U_v-modules

4.1. Integrable modules

In the next three chapters, we will explain the general theory of quantized enveloping algebras associated with "symmetrizable" Cartan matrices. However, as we have defined U_v as the quantum algebra of type $A^{(1)}_{r-1}$, we state the results only in this case.

As in the previous chapter, $(\mathfrak{h}, \Pi, \Pi^\vee)$ is the realization of the Cartan matrix of type $A^{(1)}_{r-1}$ defined in Definition 3.8, and the base field of the quantum algebra is $K = \mathbb{Q}(v)$.

LEMMA 4.1. *We can define a K-algebra homomorphism $\Delta : U_v \to U_v \otimes U_v$ by*

$$\Delta(t_i) = t_i \otimes t_i, \quad \Delta(v^d) = v^d \otimes v^d,$$
$$\Delta(e_i) = 1 \otimes e_i + e_i \otimes t_i^{-1},$$
$$\Delta(f_i) = f_i \otimes 1 + t_i \otimes f_i.$$

The proof is obtained simply by checking the defining relations. Similar to the case $U_v(\mathfrak{sl}_r)$ in Chapter 2, we can define tensor product representations of U_v using Δ.

DEFINITION 4.2. A U_v-module M is **integrable** if it satisfies the following three conditions.

(1) Set $P := \{\lambda \in \mathfrak{h}^* | \ \lambda(h_i), \lambda(d) \in \mathbb{Z}\}$. Then M has a weight space decomposition. That is, if we define the weight space M_λ, for $\lambda \in P$, by
$$M_\lambda = \{m \in M | \ t_i m = v^{\lambda(h_i)} m, \ (0 \le i < r), \ v^d m = v^{\lambda(d)} m\},$$
then we have the simultaneous eigenspace decomposition $M = \oplus_{\lambda \in P} M_\lambda$ with respect to the elements $t_0, \ldots, t_{r-1}, v^d \in U_v$.

(2) We have $\dim_K M_\lambda < \infty$.

(3) For any i, let $U_v(\mathfrak{g}_i)$ be the subalgebra of U_v generated by e_i, h_i, f_i. Then any element $m \in M$ is contained in a finite dimensional $U_v(\mathfrak{g}_i)$-submodule.

Note that (3) implies the following.

(3′) For any $m \in M$, we have $e_i^n m = 0, f_i^n m = 0$ for large enough n. That is, e_i and f_i are locally nilpotent on M for all i.

Conversely, (3) follows from (1) and (3′). In fact, since any element m of M is a linear combination of eigenvectors of t_i by (1), we may assume that m is an eigenvector of t_i. We set $M' = \sum_{N_1, N_2 \ge 0} K f_i^{N_1} e_i^{N_2} m$. Since e_i is locally nilpotent, $\sum_{N_2 \ge 0} K e_i^{N_2} m$ is finite dimensional. Since f_i is also locally nilpotent, we have that M' is finite dimensional. Finally, we have $t_i M' \subset M'$, $f_i M' \subset M'$ and we can prove that $e_i M' \subset M'$ using

$$e_i f_i^n = \frac{t_i - t_i^{-1}}{v - v^{-1}} f_i^{n-1} + f_i e_i f_i^{n-1}. \tag{4.1}$$

Therefore, M' is a $U_v(\mathfrak{g}_i)$-submodule. The formula (4.1) is proved by induction on n. To summarize, we can replace (3) by (3$'$) in the definition above.

In later chapters, we will meet important examples of an integrable module, namely, integrable highest weight modules. We list some other examples here. Example 4.3 generalizes to type $A_{r-1}^{(1)}$ for $r \geq 3$.

EXAMPLE 4.3. Let $r = 2$, and view $M = K^2 \otimes K[\xi, \xi^{-1}]$ as a U_v-module via

$$e_0 = \begin{pmatrix} 0 & 0 \\ 1 & 0 \end{pmatrix} \otimes \xi, \quad f_0 = \begin{pmatrix} 0 & 1 \\ 0 & 0 \end{pmatrix} \otimes \xi^{-1},$$

$$e_1 = \begin{pmatrix} 0 & 1 \\ 0 & 0 \end{pmatrix} \otimes 1, \quad f_1 = \begin{pmatrix} 0 & 0 \\ 1 & 0 \end{pmatrix} \otimes 1,$$

$$t_0 = \begin{pmatrix} v^{-1} & 0 \\ 0 & v \end{pmatrix} \otimes 1, \quad t_1 = \begin{pmatrix} v & 0 \\ 0 & v^{-1} \end{pmatrix} \otimes 1$$

and

$$v^d (m \otimes \xi^j) = v^j m \otimes \xi^j.$$

To show that M is a U_v-module, we check the relation

$$e_0^3 e_1 - (v^2{+}1{+}v^{-2}) e_1 e_0 e_1^2 + (v^2{+}1{+}v^{-2}) e_1^2 e_0 e_1 - e_1 e_0^3 = 0$$

as well as other defining relations, which are easy. Because of the weight space decomposition

$$M = \left(\bigoplus_{j \in \mathbb{Z}} K \begin{pmatrix} 1 \\ 0 \end{pmatrix} \otimes \xi^j \right) \bigoplus \left(\bigoplus_{j \in \mathbb{Z}} K \begin{pmatrix} 0 \\ 1 \end{pmatrix} \otimes \xi^j \right),$$

we know that M is integrable. We can check that M has no non-zero vector m that satisfies $e_i m = 0\, (i = 0, 1)$. (This implies that M is not an integrable highest weight module.)

EXAMPLE 4.4. Let M be an integrable U_v-module. If we change the action of e_i, t_i to $-e_i, -t_i$, we have another U_v-module structure. This is not integrable since the eigenvalues of t are not powers of v.

The representation theory of the quantum algebra $U_v(\mathfrak{sl}_2)$ is a basic tool for analyzing the representations of any quantum algebra. In the next section, we define the Kashiwara operators which will be used to introduce crystal bases.

4.2. The Kashiwara operators

We start with the classification of finite dimensional irreducible representations of $U_v(\mathfrak{sl}_2)$. In the following, $[k] = \frac{v^k - v^{-k}}{v - v^{-1}}$ and $[n]! = \prod_{k=1}^{n} [k]$ as before.

PROPOSITION 4.5. *Let $V_l^{\pm}$ be the $(l+1)$-dimensional representation of $U_v(\mathfrak{sl}_2)$ given as follows.*

$$V_l^{\pm} : e = \pm \begin{pmatrix} 0 & [l] & & \\ & \cdots & \cdots & \\ & & 0 & [1] \\ & & & 0 \end{pmatrix}, \quad f = \begin{pmatrix} 0 & & & \\ [1] & 0 & & \\ & \cdots & \cdots & \\ & & [l] & 0 \end{pmatrix},$$

$$t = \pm \begin{pmatrix} v^l & & & \\ & \cdots & & \\ & & \cdots & \\ & & & v^{-l} \end{pmatrix}.$$

Then $\{V_l^{\pm} | \, l = 0, 1, \dots\}$ is a complete set of pairwise non-isomorphic finite dimensional irreducible representations of $U_v(\mathfrak{sl}_2)$.

PROOF. It is easy to see that these are irreducible representations. Let M be an arbitrary finite dimensional irreducible module. We first observe that the eigenvalues of e on M are all 0. In fact, if $em = \lambda m$, then $e(t^N m) = (v^{-2N}\lambda)t^N m$ for any positive integer N. This implies that $\lambda = 0$ since the number of eigenvalues of e is finite. The same is true for f, so we have that e and f are nilpotent. We now consider $M_0 = \{m \in M \mid em = 0\}$. Since $M_0 \neq 0$ and $tM_0 \subset M_0$, we can choose $m \neq 0$ which satisfies $tm = cm$, $em = 0$, and we have $f^l m \neq 0, f^{l+1}m = 0$ for some l. Then

$$\begin{aligned} &\frac{(v - v^{-2l-1})c + (v^{-1} - v^{2l+1})c^{-1}}{(v - v^{-1})^2} f^l m \\ &= \frac{1}{(v - v^{-1})^2} f^l \left((v - v^{-2l-1})t + (v^{-1} - v^{2l+1})t^{-1}\right) m \\ &= ef^{l+1}m - f^{l+1}em = 0, \end{aligned}$$

so $c = \pm v^l$. Therefore, we have a $U_v(\mathfrak{sl}_2)$-submodule of M whose basis is given by $\{m, fm, \dots, f^{(l)}m\}$ where $f^{(i)} = \frac{f^i}{[i]!}$, and whose matrix representation using this basis is $V_l^{\pm}$. By the irreducibility of M, this $U_v(\mathfrak{sl}_2)$-submodule coincides with M. □

To show that any finite dimensional representation is completely reducible, we introduce the Casimir element of $U_v(\mathfrak{sl}_2)$.

DEFINITION 4.6. The **Casimir element** of $U_v(\mathfrak{sl}_2)$ is the element

$$C = fe + \frac{vt - v^{-1}t^{-1}}{(v - v^{-1})^2}.$$

The Casimir element belongs to the center of $U_v(\mathfrak{sl}_2)$ and has distinct eigenvalues $\pm\frac{[l+1]}{v - v^{-1}}$ on $V_l^{\pm}$.

THEOREM 4.7. *Every finite dimensional representation of $U_v(\mathfrak{sl}_2)$ is completely reducible, i.e. it is isomorphic to a direct sum of irreducible representations.*

PROOF. Let $0 \to W \xrightarrow{\iota} V \to U \to 0$ be an exact sequence. We shall show that it splits. The proof is given in the following steps.

(step 1) *If the sequence splits when $U = V_0^+$ then it splits in general.*

To prove this, let id_W be the identity operator on W and set

$$V' = \{s \in \mathrm{Hom}_K(V, W) \mid s \circ \iota \in K\mathrm{id}_W\},$$
$$W' = \{s \in \mathrm{Hom}_K(V, W) \mid s \circ \iota = 0\}.$$

Then V' and W' become $U_v(\mathfrak{sl}_2)$-modules via the following action

$$(e \cdot s)(m) = e(s(tm)) - s(etm), \quad (t \cdot s)(m) = t(s(t^{-1}m)),$$
$$(f \cdot s)(m) = f(s(m)) - t(s(t^{-1}fm)).$$

If $s \in W'$, we have

$$(e \cdot s) \circ \iota(m) = e(s(\iota(tm))) - s(\iota(etm)) = 0,$$
$$(f \cdot s) \circ \iota(m) = f(s(\iota(m))) - ts(\iota(t^{-1}fm)) = 0,$$
$$(t \cdot s) \circ \iota(m) = t(s(\iota(t^{-1}m))) = 0,$$

which shows that W' is a submodule of V' and V'/W' is a one dimensional representation. By changing the action to $e \mapsto -e$, $t \mapsto -t$ if necessary, we may assume that $V'/W' \simeq V_0^+$. Hence we are in the case where $U = V_0^+$ and, by assumption, we can choose $s \in \mathrm{Hom}_K(V, W)$ so that $V' = W' \oplus Ks$. Here $Ks \simeq V_0^+$ and in particular we have

$$e(s(tm)) = s(etm), \quad f(s(m)) = t(s(t^{-1}fm)), \quad t(s(t^{-1}m)) = s(m).$$

Thus $s \in \mathrm{Hom}_{U_v(\mathfrak{sl}_2)}(V, W)$ and by multiplying a suitable scalar, we have the desired splitting.

(step 2) *If $U = V_0^+$ then the exact sequence splits.*

This is proved by induction on the dimesion of W. If W is one dimensional, then W is isomorphic to one of $V_0^{\pm}$. Hence we can verify the statement by an explicit computation. Assume that the sequence splits for $\dim W \le l$ and that W has dimension $l+1$. If W is irreducible, then $W = V_l^{\pm}$. Since $V_l^{\pm}$ are not equivalent to V_0^+, the generalized eigenspace decomposition of V with respect to the Casimir element C gives the desired splitting. If W is not irreducible, we take a non-zero proper submodule W_0 of W and apply the induction hypothesis to the following exact sequence.

$$0 \to W/W_0 \to V/W_0 \to V_0^+ \to 0$$

Then we can choose a submodule W_1 of V which satisfies

$$W_1/W_0 \simeq V_0^+, \quad V = W + W_1, \quad W \cap W_1 = W_0.$$

We then apply the induction hypothesis to $0 \to W_0 \to W_1 \to V_0^+ \to 0$. As a result, we can choose $s \in W_1$ which satisfies $Ks \simeq V_0^+$, $W_1 = W_0 \oplus Ks$. Now it remains to prove that $V = W \oplus Ks$, which is easy. $\square$

COROLLARY 4.8. *Suppose that M is an integrable U_v-module. Then*

$$M = \bigoplus_{\lambda \in P} \left(\bigoplus_{0 \le n \le \lambda(h_i)} f_i^{(n)}(\mathrm{Ker}\, e_i \cap M_\lambda) \right).$$

PROOF. We consider sets $\{L_\alpha\}_{\alpha\in I}$ consisting of finite dimensional irreducible $U_v(\mathfrak{g}_i)$-submodules of M which have the property that the sum $\sum_{\alpha\in I} L_\alpha$ is direct. We introduce an order by inclusion of indexing sets. This makes the set of these sets into an inductively ordered set. Zorn's lemma guarantees the existence of a maximal element, which we denote by $\{L_\alpha^{\max}\}_{\alpha\in I}$. Since $\{L_\alpha^{\max}\}_{\alpha\in I}$ is maximal, every finite dimensional irreducible $U_v(\mathfrak{g}_i)$-submodule of M is contained in $\oplus_{\alpha\in I} L_\alpha^{\max}$.

Now any element $m \in M$ is in a finite dimensional $U_v(\mathfrak{g}_i)$-submodule by the integrability. Further, the complete reducibility (Theorem 4.7) implies that m is a sum of elements each of which belongs to a finite dimensional irreducible $U_v(\mathfrak{g}_i)$-submodule. Thus, we have $m \in \oplus_{\alpha\in I} L_\alpha^{\max}$. We have proved $M = \oplus_{\alpha\in I} L_\alpha^{\max}$. Recalling that each $L_\alpha^{\max}$ is isomorphic to some V_l^+ and that $\operatorname{Ker} e_i$ admits a weight space decomposition, we have the desired decomposition. ☐

By this corollary, we can define the following operators.

DEFINITION 4.9. For an integrable module M define operators $\tilde{e}_i$ and $\tilde{f}_i$ by

$$\tilde{e}_i(\sum f_i^{(n)} u_n) = \sum f_i^{(n-1)} u_n \quad (u_n \in \operatorname{Ker} e_i),$$

$$\tilde{f}_i(\sum f_i^{(n)} u_n) = \sum f_i^{(n+1)} u_n \quad (u_n \in \operatorname{Ker} e_i).$$

The operators $\tilde{e}_i$ and $\tilde{f}_i$, for $0 \le i \le r-1$, are the **Kashiwara operators**.

4.3. Crystal bases

We are now in a position to state the definition of the crystal basis, which is the theme of these lectures.

DEFINITION 4.10. Let $R = \mathbb{Q}[v]_{(v)}$ be the localization of $\mathbb{Q}[v]$ with respect to the ideal (v). For an integrable U_v-module M, we say that (L, B) is a **crystal basis** of M if the following conditions are satisfied.

(1) L is a full rank R-lattice of M, and $\tilde{e}_i L \subset L$, $\tilde{f}_i L \subset L$ for all i.
(2) B is a $\mathbb{Q}$-basis of L/vL, and $\tilde{e}_i B \subset B \cup \{0\}$, $\tilde{f}_i B \subset B \cup \{0\}$ for all i.
(3) Set $L_\lambda = M_\lambda \cap L$, $B_\lambda = (L_\lambda/vL_\lambda) \cap B$. Then both L and B have the weight space decompositions; that is,

$$L = \bigoplus_{\lambda\in P} L_\lambda, \quad B = \bigsqcup_{\lambda\in P} B_\lambda.$$

(4) For $b_1, b_2 \in B$, we have $\tilde{e}_i b_1 = b_2 \Leftrightarrow \tilde{f}_i b_2 = b_1$ for all i.

To a crystal basis, we can associate a colored oriented graph as follows.

DEFINITION 4.11. Let M be an integrable U_v-module and let (L, B) be a crystal basis of M. The **crystal graph** of (L, B) is the colored oriented graph whose vertex set is B, and whose colored edges are given by $b \xrightarrow{i} b'$ whenever $\tilde{f}_i b = b'$.

The following two definitions are also important.

DEFINITION 4.12. Let M_1 and M_2 be integrable U_v-modules and let (L_1, B_1) and (L_2, B_2) be crystal bases of M_1 and M_2 respectively. We say that $(L_1, B_1) \simeq (L_2, B_2)$ if there exists an isomorphism of U_v-modules $\phi : M_1 \simeq M_2$ which induces an isomorphism of R-lattices $\phi : L_1 \simeq L_2$ such that $\bar{\phi}(B_1) = B_2$ under the induced isomorphism of the vector spaces $\bar{\phi} : L_1/vL_1 \simeq L_2/vL_2$.

DEFINITION 4.13. Let M be an integrable U_v-module and let $M = M_1 \oplus M_2$ be a direct sum decomposition of M into U_v-submodules. Assume that we are given crystal bases (L_1, B_1) and (L_2, B_2) of M_1 and M_2.

We say that $(L, B) = (L_1, B_1) \oplus (L_2, B_2)$ if

$$L = L_1 \oplus L_2,\ B = B_1 \sqcup B_2 \subset L_1/vL_1 \oplus L_2/vL_2 = L/vL.$$

Under these circumstances, (L, B) is a crystal basis of M. The proof is straightforward. The following lemma follows easily from definitions.

LEMMA 4.14. *Let M be an integrable U_v-module and let (L, B) be a crystal basis of M. Assume that there exists a direct sum decomposition $M = \oplus M_x$ of M into U_v-submodules such that $L = \oplus L_x$ where $L_x = L \cap M_x$, and $B = \sqcup B_x$ where $B_x = B \cap (L_x/vL_x)$. Then (L_x, B_x) is a crystal basis of M_x and $(L, B) = \oplus(L_x, B_x)$.*

EXERCISE 4.15. Let M_0 be an integrable module, (L_0, B_0) a crystal basis of M_0. Show that (L, B) where $L = L_0 \oplus L_0$ and $B = \{(b, \pm b) \mid b \in B_0\}$ is a crystal basis of $M = M_0 \oplus M_0$. Show also that $(L, B) \simeq (L_0, B_0)^{\oplus 2}$.

CHAPTER 5

The tensor product of crystals

5.1. Basics of crystal bases

We start with $U_v(\mathfrak{sl}_2)$ which lays the foundation for the general case. We use the same notation as in the previous chapter: $R = \mathbb{Q}[v]_{(v)}, K = \mathbb{Q}(v)$, $[k] = \frac{v^k - v^{-k}}{v - v^{-1}}$ and $[n]! = \prod_{k=1}^n [k]$.

Let $V_l = V_l^+ = \oplus_{i=0}^l K u_i$ be the $(l+1)$-dimensional representation of $U_v(\mathfrak{sl}_2)$ defined in Proposition 4.5: the action of the generators on the basis $\{u_i\}_{0 \le i \le l}$ is given by

$$e \cdot u_i = [l+1-i] u_{i-1}, \quad f \cdot u_i = [i+1] u_{i+1}, \quad t \cdot u_i = v^{l-2i} u_i,$$

and $f^{(k)} u_0 = u_k$ where $f^{(k)} = \frac{f^k}{[k]!}$.

LEMMA 5.1. *Define an R-lattice L_l of V_l and a basis B_l of L_l / vL_l by*

$$L_l = \bigoplus_{0 \le i \le l} R u_i, \quad B_l = \{u_i \pmod{vL_l}\}_{0 \le i \le l}.$$

Then (L_l, B_l) is a crystal basis of V_l, and any crystal basis of V_l coincides with (L_l, B_l) up to a scalar multiple.

PROOF. It is clear that (L_l, B_l) is a crystal basis. Take a crystal basis (L, B) of V_l. Multiplying by a nonzero constant, we may assume that $L \cap K u_0 = R u_0$ and $u_0 \pmod{vL} \in B$. Since $\tilde{e}$ and $\tilde{f}$ give linear isomorphisms $R u_i \simeq R u_{i+1}$ $(0 \le i \le l-1)$, we have $L = \sum R u_i$, $B = B_l$. □

PROPOSITION 5.2. *Let M be an integrable U_v-module and (L, B) a crystal basis of M. Then we have the following.*

1. *For $u \in L_\lambda$, we write $u = \sum_{n=0}^N f_i^{(n)} u_n$ ($u_n \in \operatorname{Ker} e_i \cap M_{\lambda + n\alpha_i}$). Then we have $u_n \in L$ $(0 \le n \le N)$.*
2. *If $u \in L_\lambda$ satisfies $u \pmod{vL} \in B$, then there exists $u_{n_0} \in \operatorname{Ker} e_i \cap L_{\lambda + n_0 \alpha_i}$ such that $u_{n_0} \pmod{vL} \in B$ and $u \equiv f_i^{(n_0)} u_{n_0} \pmod{vL}$.*
3. *We have* $L = \bigoplus_{\lambda \in P} \left(\bigoplus_{n=0}^{\lambda(h_i)} f_i^{(n)} (\operatorname{Ker} e_i \cap L_\lambda) \right)$.
4. *Let $M|_{U_v(\mathfrak{g}_i)}$ be the module M which is considered as a $U_v(\mathfrak{g}_i)$-module. The pair (L, B) which is considered as a crystal basis of $M|_{U_v(\mathfrak{g}_i)}$ is denoted by $(L, B)|_{U_v(\mathfrak{g}_i)}$. Then $(L, B)|_{U_v(\mathfrak{g}_i)}$ admits a direct sum decomposition with each direct summand isomorphic to one of the (L_l, B_l) defined in Lemma 5.1 as a crystal basis. Here we identify $U_v(\mathfrak{g}_i)$ with $U_v(\mathfrak{sl}_2)$ via*

$$e_i \mapsto e, \quad f_i \mapsto f, \quad t_i \mapsto t.$$

PROOF. (1) This is proved by induction on N. The case $N = 0$ is obvious. Assume that we have proved the assertion for the cases $0, \ldots, N-1$. We apply the induction hypothesis to $\tilde{e}_i u \in L$: since $\tilde{e}_i u = \sum_{n=1}^{N} f_i^{(n-1)} u_n$, we have $u_1, \ldots, u_N \in L$. Thus we also have $u_0 = u - \sum_{n=1}^{N} f_i^{(n)} u_i \in L$.
(2) It is enough to prove the following assertion. We argue by induction on N.

Assertion *Assume that $u \in L_\lambda$ and $u \pmod{vL} \in B_\lambda$, and write*

$$u = \sum_{n=0}^{N} f_i^{(n)} u_n \quad (u_n \in \operatorname{Ker} e_i \cap M_{\lambda + n\alpha_i}).$$

Then we have $u \equiv f_i^{(n_0)} u_{n_0} \pmod{vL}$ for some n_0.

To prove this, first assume that $\tilde{e}_i u = \sum_{n=1}^{N} f_i^{(n-1)} u_n \in vL$. By applying (1) to $v^{-1}\tilde{e}_i u \in L$, we get $u_0 \in L$, $u_n \in vL$ $(1 \le n \le N)$. Thus $u \equiv u_0 \pmod{vL}$ in this case. Next assume that $\tilde{e}_i u \notin vL$. Recalling property (4) of the crystal basis, we have $\tilde{e}_i u \pmod{vL} \in B$. Hence there exists n_0 such that we have $\tilde{e}_i u \equiv f_i^{(n_0 - 1)} u_{n_0} \pmod{vL}$ by the induction hypothesis. Applying $\tilde{f}_i$ to both sides, we obtain the desired equation since $u \pmod{vL} = \tilde{f}_i \tilde{e}_i (u \pmod{vL})$ and $\tilde{f}_i(vL) \subset vL$.
(3) Take an element $u = \sum_{n=0}^{N} f_i^{(n)} u_n$ of L_λ. Since $u \in L$ implies $u_n \in L$ by (1), we have the following inclusion.

$$L \subset \sum_{\lambda \in P} \sum_{n \ge 0} f_i^{(n)} (\operatorname{Ker} e_i \cap L_\lambda) = \sum_{l \ge 0} \bigoplus_{n=0}^{l} f_i^{(n)} \left(\bigoplus_{\substack{\lambda \in P \\ \lambda(h_i) = l}} (\operatorname{Ker} e_i \cap L_\lambda) \right).$$

Since the eigenvalues of the Casimir element of $U_v(\mathfrak{g}_i)$ distinguishes V_l, we know that the sum over l is direct. Thus, we have

$$L \subset \bigoplus_{\lambda \in P} \left(\bigoplus_{n=0}^{\lambda(h_i)} f_i^{(n)} (\operatorname{Ker} e_i \cap L_\lambda) \right).$$

The opposite inclusion is obvious because $\tilde{f}_i L \subset L$ and $\operatorname{Ker} e_i \cap L_\lambda \subset L$ imply that $f_i^{(n)} (\operatorname{Ker} e_i \cap L_\lambda) \subset L$.
(4) Now $(\operatorname{Ker} e_i \cap L_\lambda)/v(\operatorname{Ker} e_i \cap L_\lambda)$ may be considered as a subspace of L/vL by (3), so it makes sense to define

$$B_\lambda^{hw} := B \cap ((\operatorname{Ker} e_i \cap L_\lambda)/v(\operatorname{Ker} e_i \cap L_\lambda)).$$

We have $B \subset \sqcup_{\lambda \in P} \{ \tilde{f}_i^n b \mid b \in B_\lambda^{hw}, 0 \le n \le \lambda(h_i) \}$ by (2). Property (4) gives the opposite inclusion.

For each $b \in B_\lambda^{hw}$, we take a vector $m_b \in \operatorname{Ker} e_i \cap L_\lambda$ so that its projection to L/vL coincides with b. We multiply these vectors by $f_i^{(n)}$ $(0 \le n \le \lambda(h_i))$ and take the union over $\lambda \in P$ of all such vectors. By Nakayama's lemma, this set is an R-basis of L. Thus

$$L = \bigoplus_{\lambda \in P} \bigoplus_{b \in B_\lambda^{hw}} \left(\sum_{n=0}^{\lambda(h_i)} R f_i^{(n)} m_b \right).$$

On the other hand, if $b \in B_\lambda^{hw}$ then $\sum_{n=0}^{\lambda(h_i)} K f_i^{(n)} m_b \simeq V_{\lambda(h_i)}$ and

$$\left(\sum_{n=0}^{\lambda(h_i)} R f_i^{(n)} m_b, \{ f_i^{(n)} m_b \pmod{vL} \}_{0 \le n \le \lambda(h_i)} \right)$$

is isomorphic to $(L_{\lambda(h_i)}, B_{\lambda(h_i)})$ as a crystal basis. Therefore, the direct sum of the $U_v(\mathfrak{g}_i)$-submodules generated by all such m_b has the desired property. □

Note that Proposition 5.2 is valid for $r = 1$, which is the case where $U_v = U_v(\mathfrak{sl}_2)$. We will use this fact later.

LEMMA 5.3. *The following hold in* $U_v(\mathfrak{sl}_2)$.

(1) $\Delta(f^{(n)}) = \sum_{k=0}^{n} v^{-k(n-k)} f^{(k)} t^{n-k} \otimes f^{(n-k)}$.

(2) *Let* $V = V_1 = Ku_+ \oplus Ku_-$ *and* $V_l = Ku_0 \oplus \cdots \oplus Ku_l$ *be the irreducible* $U_v(\mathfrak{sl}_2)$*-modules defined in Lemma 5.1. Then* $V_l \otimes_K V \simeq V_{l+1} \oplus V_{l-1}$. *Further, if we regard* V_{l+1} *and* V_{l-1} *as* $U_v(\mathfrak{sl}_2)$*-submodules of* $V_l \otimes_K V$ *then* $\operatorname{Ker} e \cap V_{l+1}$ *and* $\operatorname{Ker} e \cap V_{l-1}$ *are spanned by* $u_0 \otimes u_+$ *and* $\left(\frac{1-v^{2l}}{1-v^2}\right) u_0 \otimes u_- - v^l u_1 \otimes u_+$ *respectively.*

(3) *Define* (L_l, B_l) *and* (L_1, B_1) *by*

$$L_l = \bigoplus_{i=0}^{l} Ru_i, \quad B_l = \{ u_i \pmod{vL_l} \},$$
$$L_1 = Ru_+ \oplus Ru_-, \quad B_1 = \{ u_\pm \pmod{vL_1} \}.$$

Then $(L_l \otimes_R L_1, B_l \times B_1)$ *is a crystal basis of* $V_l \otimes_K V$. *The action of* $\tilde{f}$ *on* $B_l \times B_1$ *is given by the following formulas.*

$$\tilde{f}(u_i \otimes u_+) \equiv \begin{cases} u_{i+1} \otimes u_+ & (0 \le i < l) \\ u_l \otimes u_- & (i = l) \end{cases} \pmod{vL_l \otimes L_1}$$

$$\tilde{f}(u_i \otimes u_-) \equiv \begin{cases} u_{i+1} \otimes u_- & (0 \le i < l-1) \\ 0 & (i = l-1, l) \end{cases} \pmod{vL_l \otimes L_1}$$

In other words, the crystal graph of this crystal basis is given as follows.

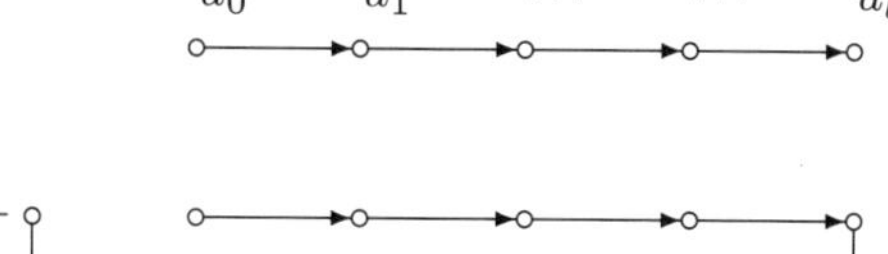

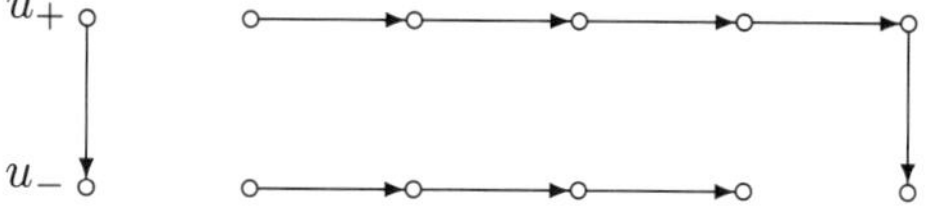

Here, the nodes represent the elements $u_i \otimes u_\pm \pmod{vL_l \otimes L_1}$ *of* $B_l \times B_1$.

PROOF. (1) This is proved by induction on n.
(2) We define w'_k $(0 \le k \le l+1)$, w''_k $(0 \le k \le l-1)$ as follows. (We understand that $u_{l+1} = 0, u_{-1} = 0$.)

$$w'_k = u_k \otimes u_+ + v^{l-k+1} u_{k-1} \otimes u_-,$$

$$w''_k = \frac{1 - v^{2l-2k}}{1 - v^2} u_k \otimes u_- - (v^{l-k} + v^{l-k+2} + \cdots + v^{l+k}) u_{k+1} \otimes u_+.$$

Then it is easy to see that $ew'_0 = 0, ew''_0 = 0$ and $f^{(k)}w'_0 = w'_k, f^{(k)}w''_0 = w''_k$ where $w'_{l+2} = 0, w''_l = 0$. Thus, we have an explicit direct sum decomposition of $V_l \otimes V$ into $U_v(\mathfrak{sl}_2)$-submodules $\sum_{k=0}^{l+1} Kw'_k$ and $\sum_{k=0}^{l-1} Kw''_k$, and the result follows.
(3) Since $w'_k, w''_k \in L_l \otimes L_1$, if we describe this basis in terms of the monomial basis $u_k \otimes u_\pm$, then the entries of the transition matrix are in R. Further, we have the following equalities in $(L_l \otimes L_1)/v(L_l \otimes L_1)$.

$$w'_k \equiv \begin{cases} u_k \otimes u_+ & (0 \le k \le l) \\ u_l \otimes u_- & (k = l+1) \end{cases}, \quad w''_k \equiv u_k \otimes u_- \ (0 \le k \le l-1).$$

In particular, the inverse of the transition matrix has entries in R. Hence, if we set

$$L' = \bigoplus_{k=0}^{l+1} Rw'_k, \quad L'' = \bigoplus_{k=0}^{l-1} Rw''_k,$$

then we have $L_l \otimes L_1 = L' \oplus L''$. If we further set

$$\begin{aligned} B' &= \{u_k \otimes u_+ \ (\mathrm{mod}\ vL_l \otimes L_1)\}_{0 \le k \le l} \cup \{u_l \otimes u_- \ (\mathrm{mod}\ vL_l \otimes L_1)\}, \\ B'' &= \{u_k \otimes u_- \ (\mathrm{mod}\ vL_l \otimes L_1)\}_{0 \le k \le l-1}, \end{aligned}$$

then we have $B_l \times B_1 = B' \sqcup B''$, and $(L', B'), (L'', B'')$ are crystal bases of V_{l+1} and V_{l-1} by Lemma 5.1. Hence $(L_l \otimes L_1, B_l \times B_1) = (L', B') \oplus (L'', B'')$ is a crystal basis of $V_l \otimes V_1$. Lemma 5.1 also shows that the action of $\tilde{f}$ on $B_l \times B_1$ is as stated. □

PROPOSITION 5.4. *Let l_1 and l_2 be positive integers.*

(1) $V_{l_1} \otimes_K V_{l_2} \simeq V_{l_1+l_2} \oplus V_{l_1+l_2-2} \oplus \cdots \oplus V_{|l_1-l_2|}$ *as a* $U_v(\mathfrak{sl}_2)$*-module.*
(2) *Let (L, B) be a crystal basis of $V_{l_1} \otimes_K V_{l_2}$. For $|l_1 - l_2| \le i \le l_1 + l_2$, we regard V_i as $U_v(\mathfrak{sl}_2)$-submodules of $V_{l_1} \otimes_K V_{l_2}$ and set $L^i = L \cap V_i$. Then $L = \oplus L^i$. If we further set $B^i = B \cap (L^i/vL^i)$, then (L^i, B^i) is a crystal basis of V_i and $(L, B) = \bigoplus (L^i, B^i)$.*
(3) *Let (L_l, B_l) be as in Lemma 5.1. Then $(L_{l_1} \otimes_R L_{l_2}, B_{l_1} \times B_{l_2})$ is a crystal basis of $V_{l_1} \otimes_K V_{l_2}$ and the action of $\tilde{f}$ on $b_1 \otimes b_2 \in B_{l_1} \times B_{l_2}$ is given by the following graph. The b_i-axis $(i = 1, 2)$ displays the crystal graph of (L_{l_i}, B_{l_i}).*

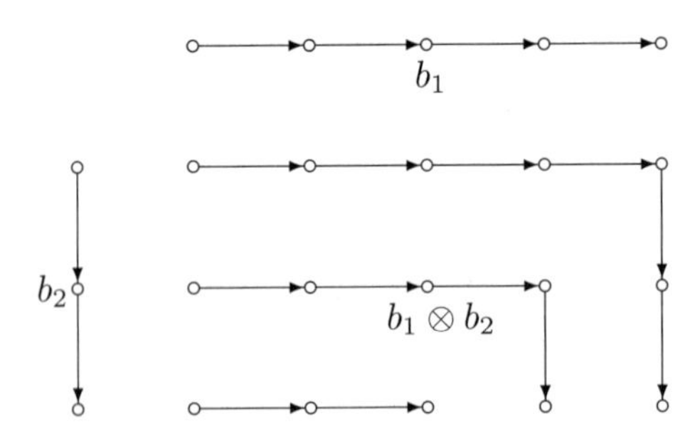

In other words, if we define $\varphi(b)$ and $\epsilon(b)$ by

$$\varphi(b) = \max\{k \in \mathbb{Z}_{\ge 0} | \tilde{f}^k b \ne 0\}, \ \epsilon(b) = \max\{k \in \mathbb{Z}_{\ge 0} | \tilde{e}^k b \ne 0\},$$

then

$$\tilde{f}(b_1 \otimes b_2) = \begin{cases} \tilde{f}b_1 \otimes b_2 & (\varphi(b_1) > \epsilon(b_2)) \\ b_1 \otimes \tilde{f}b_2 & (\varphi(b_1) \le \epsilon(b_2)) \end{cases}.$$

PROOF. (1) Take the bases of V_{l_1} and V_{l_2} given in Lemma 5.1, and denote these bases by $\{u_k\}_{0\le k\le l_1}$ and $\{v_k\}_{0\le k\le l_2}$ respectively. We consider the following vectors for $0 \le l \le \min(l_1, l_2)$.

$$w_{l_1+l_2-2l} = \sum_{0\le j\le l} (-1)^j v^{(l-j)(l_2-l+1-j)}[l_1-l+j]!\,[l_2-j]!\,u_{l-j}\otimes v_j.$$

These vectors are in $\operatorname{Ker} e \subset V_{l_1} \otimes V_{l_2}$. If we identify the $U_v(\mathfrak{s}l_2)$-submodules generated by $w_{l_1+l_2-2l}$ with $V_{l_1+l_2-2l}$, then the sum is direct since the eigenvalues of the Casimir element on these submodules are mutually distinct. Comparing the dimensions of $V_{l_1} \otimes_K V_{l_2}$ and $\oplus_{0\le l\le\min(l_1,l_2)} V_{l_1+l_2-2l}$, we have the result.
(2) Since Proposition 5.2(4) holds for $U_v(\mathfrak{s}l_2)$, there exists a direct sum decomposition of $V_{l_1} \otimes V_{l_2}$ into irreducible $U_v(\mathfrak{s}l_2)$-submodules such that (L, B) is the direct sum of crystal bases of these submodules. Since each isotropic component is irreducible by (1), these irreducible submodules are uniquely determined, and thus the crystal bases must coincide with (L^i, B^i).
(3) This is proved by induction on l_2. The case $l_2 = 1$ holds by Lemma 5.3. Assume that the assertion is valid up to l_2. Using $(\Delta \otimes \mathrm{id}) \circ \Delta = (\mathrm{id} \otimes \Delta) \circ \Delta$, we make $V_{l_1} \otimes V_{l_2} \otimes V$ into a $U_v(\mathfrak{s}l_2)$-module.

First we consider $(L_{l_1} \otimes L_{l_2}, B_{l_1} \times B_{l_2}) \otimes (L_1, B_1)$. By the induction hypothesis, $(L_{l_1}, B_{l_1}) \otimes (L_{l_2}, B_{l_2})$ is a crystal basis of $V_{l_1} \otimes V_{l_2}$. By (2), it decomposes into the direct sum of (L^i, B^i) ($|l_1 - l_2| \le i \le l_1 + l_2$), and each (L^i, B^i) is isomorphic to (L_i, B_i) as a crystal basis by Lemma 5.1. Hence, Lemma 5.3(3) implies that

$$(L, B) = (L_{l_1} \otimes L_{l_2} \otimes L_1, B_{l_1} \times B_{l_2} \times B_1)$$

is a crystal basis. By the induction hypothesis, and Lemma 5.3(3) applied to $(L_i, B_i) \otimes (L_1, B_1)$, the action of $\tilde{f}$ on B is given by the following diagram.

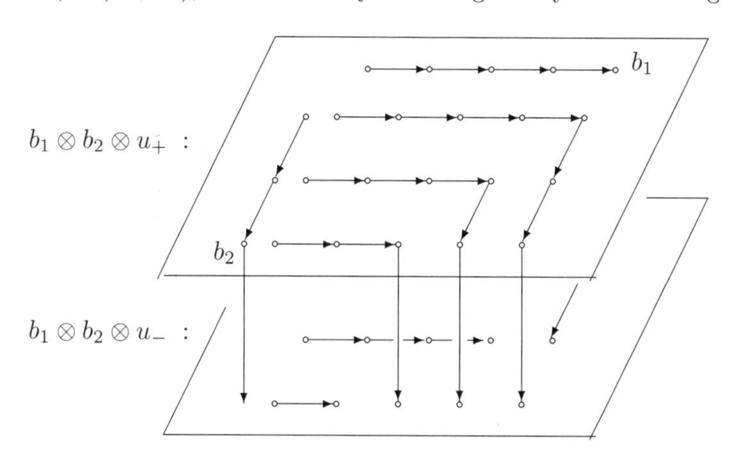

Here, the elements of $B_{l_1} \times B_{l_2} \times \{u_+ \pmod{vL_1}\}$ correspond to the nodes in the upper plane and the elements of $B_{l_1} \times B_{l_2} \times \{u_- \pmod{vL_1}\}$ correspond to the nodes in the lower plane. We also recall that the oriented edge $b \to b'$ means $\tilde{f}b = b'$.

Next, we view this crystal basis as $(L_{l_1}, B_{l_1}) \otimes (L_{l_2} \otimes L_1, B_{l_2} \times B_1)$ and apply Lemma 5.3 to $V_{l_2} \otimes V$. We claim that $V_{l_1} \otimes V_{l_2} \otimes V$ is a direct sum of two $U_v(\mathfrak{s}l_2)$-submodules $V' \simeq V_{l_1} \otimes V_{l_2+1}$ and $V'' \simeq V_{l_1} \otimes V_{l_2-1}$ such that each of $L' \simeq L_{l_1} \otimes L_{l_2+1}$ and $L'' \simeq L_{l_1} \otimes L_{l_2-1}$ is an R-lattice of V' and V'' respectively, and that $L = L' \oplus L''$.

To see this, let w'_k, w''_k be the vectors defined in the proof of Lemma 5.3 and define L' and L'' by

$$L' = \bigoplus_{k=0}^{l_2+1} L_{l_1} \otimes w'_k, \quad L'' = \bigoplus_{k=0}^{l_2-1} L_{l_1} \otimes w''_k.$$

We identify $\{w'_k (\text{mod } vL_{l_2} \otimes L_1))\}$ and $\{w''_k (\text{mod } vL_{l_2} \otimes L_1))\}$ with B_{l_2+1} and B_{l_2-1} respectively, and set $B' = B_{l_1} \times B_{l_2+1}$ and $B'' = B_{l_1} \times B_{l_2-1}$. Then we have $B = B' \sqcup B''$. Therefore, Lemma 4.14 implies that (L', B') and (L'', B'') are crystal bases and $(L, B) = (L', B') \oplus (L'', B'')$. In particular, $(L_{l_1} \otimes L_{l_2+1}, B_{l_1} \times B_{l_2+1})$ is a crystal basis of $V_{l_1} \otimes V_{l_2+1}$.

Finally, we look at the action of $\tilde{f}$ on B'. Since the action on B is explicitly described by the crystal graph (the three dimensional diagram above), it is enough to read off the decomposition $B_{l_2} \times B_1 = B_{l_2+1} \sqcup B_{l_2-1}$ from the diagram. Note that the decomposition is induced by the following decomposition.

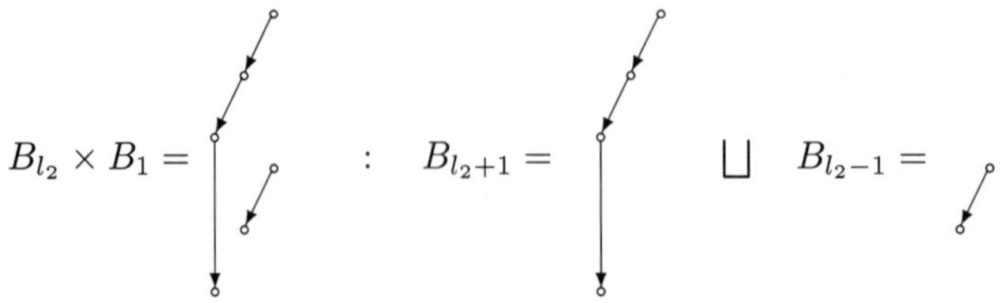

$B_{l_2} \times B_1 =$: $B_{l_2+1} =$ $\bigsqcup$ $B_{l_2-1} =$

We pick up the nodes corresponding to $B_{l_1} \times B_{l_2+1}$, and put them on a plane. Then this is the same graph as the graph for the case $l_2 + 1$. □

5.2. Tensor products of crystal bases

Based on the results proved in the previous section, we may prove the following theorem.

THEOREM 5.5 ([**Kashiwara**, Theorem 1]). *Suppose that M_i $(i = 1, 2)$ are integrable U_v-modules and let (L_i, B_i) be crystal bases of these modules. Then we have the following.*

(1) *Set $L = L_1 \otimes_R L_2$ and $B = B_1 \times B_2 = \{b_1 \otimes b_2 \ (\text{mod } vL)\}$. Then (L, B) is a crystal basis of $M_1 \otimes_K M_2$.*
(2) *Set $\varphi_i(b) = \max\{k \in \mathbb{Z}_{\geq 0} | \tilde{f}_i^k b \neq 0\}$ and $\epsilon_i(b) = \max\{k \in \mathbb{Z}_{\geq 0} | \tilde{e}_i^k b \neq 0\}$. Then the action of $\tilde{f}_i$ on B is given by*

$$\tilde{f}_i(b_1 \otimes b_2) = \begin{cases} \tilde{f}_i b_1 \otimes b_2 & (\varphi_i(b_1) > \epsilon_i(b_2)) \\ b_1 \otimes \tilde{f}_i b_2 & (\varphi_i(b_1) \leq \epsilon_i(b_2)) \end{cases}.$$

In other words, the crystal graph is characterized by the following graph. Here, the horizontal axis is a connected component of the crystal graph of (L_1, B_1) with respect to the action of $\tilde{f}_i$, and the vertical axis is a connected component of the crystal graph of (L_2, B_2) with respect to the action of $\tilde{f}_i$.

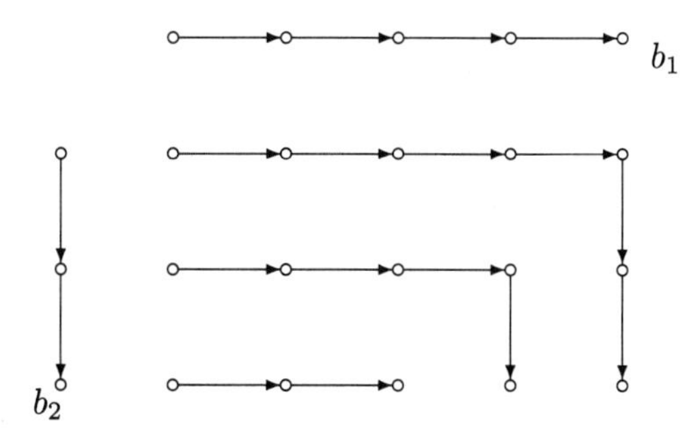

PROOF. (1) It is obvious that (L, B) has a weight space decomposition and that B is a basis of L/vL. Thus, it remains to show that (i) L is stable under the action of $\tilde{e}_i$ and $\tilde{f}_i$, (ii) $\tilde{e}_i B \subset B \cup \{0\}$, (iii) $\tilde{f}_i B \subset B \cup \{0\}$, (iv) $\tilde{e}_i b \in B$ implies $b = \tilde{f}_i \tilde{e}_i b$, and (v) $\tilde{f}_i b \in B$ implies $b = \tilde{e}_i \tilde{f}_i b$. By Proposition 5.2(4), it is enough to prove (i)-(v) in the case of $U_v(\mathfrak{sl}_2)$; they are already proved in Proposition 5.4(3).

(2) If we set

$$B_1^{hw} = B_1 \cap (L_1 \cap \text{Ker } e_i / vL_1 \cap \text{Ker } e_i),$$
$$B_2^{hw} = B_2 \cap (L_2 \cap \text{Ker } e_i / vL_2 \cap \text{Ker } e_i),$$

then by the proof of Proposition 5.2(4) we have that

$$B_1 = \bigsqcup_{\lambda \in P} \bigsqcup_{n=0}^{\lambda(h_i)} \tilde{f}_i^n (B_1^{hw})_\lambda, \quad B_2 = \bigsqcup_{\lambda \in P} \bigsqcup_{n=0}^{\lambda(h_i)} \tilde{f}_i^n (B_2^{hw})_\lambda.$$

Hence, if we consider M_1, M_2 as $U_v(\mathfrak{g}_i)$-modules then the connected components of B_1 and B_2 with respect to the action of $\tilde{f}_i$ have the form $\{\tilde{f}_i^n b\}_{0 \le n \le \lambda(h_i)}$, where b is an element of either $(B_1^{hw})_\lambda$ or $(B_2^{hw})_\lambda$. Therefore, Proposition 5.4(3) proves the assertion. $\square$

CHAPTER 6

Crystal bases of U_v^-

6.1. Triangular decomposition of U_v

In this section, we review the structure theorem which is called the **triangular decomposition** of U_v. For simplicity, we only consider the quantum algebra of type $A_{r-1}^{(1)}$ with $r \geq 3$.

DEFINITION 6.1. We denote by U_v^+, U_v^- and U_v^0 the K-subalgebras of U_v generated by $\{e_i\}_{0 \leq i \leq r-1}$, $\{f_i\}_{0 \leq i \leq r-1}$, and $\{v^{\pm d}\} \cup \{t_i^{\pm 1}\}_{0 \leq i \leq r-1}$, respectively.

THEOREM 6.2. *Let U_v^+, U_v^- and U_v^0 be as above. Then we have the following.*

(1) *There is an isomorphism of vector spaces $U_v^- \otimes_K U_v^0 \otimes_K U_v^+ \simeq U_v$ given by the product map: $u^- \otimes u^0 \otimes u^+ \mapsto u^- u^0 u^+$.*
(2) *We denote by I_+ the two-sided ideal of the tensor algebra $T(\oplus K e_i)$ generated by*

$$\begin{cases} e_i^2 e_j - (v + v^{-1}) e_i e_j e_i + e_j e_i^2 & (i - j \equiv \pm 1) \\ e_i e_j - e_j e_i & (otherwise) \end{cases},$$

and by I_- the two-sided ideal of the tensor algebra $T(\oplus K f_i)$ generated by

$$\begin{cases} f_i^2 f_j - (v + v^{-1}) f_i f_j f_i + f_j f_i^2 & (i - j \equiv \pm 1) \\ f_i f_j - f_j f_i & (otherwise) \end{cases}.$$

Then, under the obvious maps, we have the following isomorphisms of vector spaces.

$$T(\oplus_{i=0}^{r-1} K e_i)/I_+ \simeq U_v^+, \quad T(\oplus_{i=0}^{r-1} K f_i)/I_- \simeq U_v^-.$$

Further, U_v^0 is isomorphic to a Laurent polynomial ring:

$$U_v^0 \simeq K[t_0^{\pm 1}, \ldots, t_{r-1}^{\pm 1}, v^{\pm d}].$$

PROOF. By the defining relations of U_v, we know that $U_v^- U_v^0 U_v^+$ is stable under left multiplication by e_i, f_i, t_i and v^d. Hence we have $U_v = U_v^- U_v^0 U_v^+$. Let $P = K[T_0^{\pm 1}, \ldots, T_{r-1}^{\pm 1}, T_r^{\pm 1}]$ be a Laurent polynomial ring and let $\oplus_{i=0}^{r-1} K F_i$ and $\oplus_{i=0}^{r-1} K E_i$ be r-dimensional vector spaces.

(step 1) *We inductively define operators e_i, t_i, f_i, for $0 \leq i \leq r-1$, and v^d on*

$$T(\oplus_{i=0}^{r-1} K F_i) \otimes P \otimes T(\oplus_{i=0}^{r-1} K E_i)$$

as follows. In the following, $T_0^{m_0} \cdots T_{r-1}^{m_{r-1}} T_r^m$ is abbreviated by t.

$$\begin{cases} e_i(1\otimes t\otimes y) = v^{-\sum_{j=0}^{r-1} m_j\alpha_i(h_j)-m\alpha_i(d)} 1\otimes t\otimes E_i y, \\ t_i(1\otimes t\otimes y) = 1\otimes T_i t\otimes y, \quad v^d(1\otimes t\otimes y) = 1\otimes T_r t\otimes y, \end{cases}$$

$$f_i(x\otimes t\otimes y) = (F_i x)\otimes t\otimes y,$$

$$\begin{cases} e_i(F_j x\otimes t\otimes y) = \frac{\delta_{ij}}{v-v^{-1}}(t_i - t_i^{-1})(x\otimes t\otimes y) + f_j e_i(x\otimes t\otimes y), \\ t_i(F_j x\otimes t\otimes y) = v^{-\alpha_j(h_i)} f_j t_i(x\otimes t\otimes y), \\ v^d(F_j x\otimes t\otimes y) = v^{-\alpha_j(d)} f_j v^d(x\otimes t\otimes y). \end{cases}$$

Then all of the defining relations of U_v hold except for the last four.

The relations $t_i f_j = v^{-\alpha_j(h_i)} f_j t_i$, $v^d f_j = v^{-\delta_{0j}} f_j v^d$ and $[e_i, f_j] = \delta_{ij}\frac{t_i - t_i^{-1}}{v-v^{-1}}$ are clear from the definitions. The relations $t_i t_j^{\pm 1} = t_j^{\pm 1} t_i$, $v^d t_j^{\pm 1} = t_j^{\pm 1} v^d$ and $t_i e_j = v^{\alpha_j(h_i)} e_j t_i$, $v^d e_j = v^{\delta_{0j}} e_j v^d$ are proved by induction.

(step 2) *We define an operator θ_{ij}^- on $T(\oplus_{i=0}^{r-1} KF_i)\otimes P\otimes T(\oplus_{i=0}^{r-1} KE_i)$ by*

$$\theta_{ij}^- = \begin{cases} f_i^2 f_j - (v+v^{-1}) f_i f_j f_i + f_j f_i^2 & (i-j\equiv \pm 1) \\ f_i f_j - f_j f_i & (\mathit{otherwise}) \end{cases}.$$

Then we have $e_k\theta_{ij}^- = \theta_{ij}^- e_k$ for all k. Defining θ_{ij}^+ similarly, we have $f_k\theta_{ij}^+ = \theta_{ij}^+ f_k$ for all k.

Since the commutation relation of e_i and f_j was proved in (step 1), this follows by an explicit computation.

(step 3) *Define $I_\pm$ using E_i and F_i $(0\le i\le r-1)$. Then*

$$I_-\otimes P\otimes T(\oplus KE_i) + T(\oplus KF_i)\otimes P\otimes I_+$$

is stable under the operators e_i, t_i, v^d, f_i defined in (step 1).

This is obvious for t_i, v^d, f_i. For e_i, it follows from (step 2).

(step 4) *$(T(\oplus KF_i)/I_-)\otimes P\otimes (T(\oplus KE_i)/I_+)$ is a U_v-module.*

We have well-defined operators e_i, t_i, v^d, f_i on this space by (step 3). From the definitions, the element θ_{ij}^- acts as 0.

Because we have $\theta_{ij}^+(1\otimes t\otimes y) = 0$ and $\theta_{ij}^+(F_k x\otimes t\otimes y) = f_k\theta_{ij}^+(x\otimes t\otimes y)$ on this space, θ_{ij}^+ also acts as 0. Hence (step 4) follows.

(step 5) *We have an isomorphism of vector spaces*

$$U_v \simeq (T(\oplus KF_i)/I_-)\otimes P\otimes (T(\oplus KE_i)/I_+).$$

For elements $u^-\in T(\oplus KF_i)/I_-$, $u^0\in P$ and $u^+\in T(\oplus KE_i)/I_+$, let $\bar u^-\in U^-$, $\bar u^0\in U^0$ and $\bar u^+\in U^+$ be the images of u^-, u^0 and u^+ under the maps

$$T(\oplus KF_i)/I_-\to U^-,\quad P\to U^0,\quad T(\oplus KE_i)/I_+\to U^-,$$

respectively, where the maps are defined by $F_i \mapsto f_i$, $T_i \mapsto t_i$, for $0 \leq i \leq r-1$, $T_r \mapsto v^d$, and $E_i \mapsto e_i$. Then we have

$$\begin{aligned} \bar{u}^- \bar{u}^0 \bar{u}^+ (1 \otimes 1 \otimes 1) &= \bar{u}^- \bar{u}^0 (1 \otimes 1 \otimes u^+) \\ &= \bar{u}^- (1 \otimes u^0 \otimes u^+) = u^- \otimes u^0 \otimes u^+. \end{aligned} \tag{6.1}$$

Let $a \in U_v$ act on $1 := 1 \otimes 1 \otimes 1$. Then the map $a \mapsto a \cdot 1$ is a surjective map from U_v to $(T(\oplus K F_i)/I_-) \otimes P \otimes (T(\oplus K E_i)/I_+)$ by $U_v = U_v^- U_v^0 U_v^+$ and the equality (6.1). We compose this map with the natural surjection

$$(T(\oplus K F_i)/I_-) \otimes P \otimes (T(\oplus K E_i)/I_+) \to U_v^- \otimes U_v^0 \otimes U_v^+ \to U_v.$$

This composition is the identity map on $(T(\oplus K F_i)/I_-) \otimes P \otimes (T(\oplus K E_i)/I_+)$, and thus we have the desired isomorphisms in (1) and (2). □

The root space decomposition for the quantum algebra is as follows. Unlike the case of Lie algebras, we have to define the root spaces in $U_v^{\pm}$ separately for the cases of enveloping algebras and quantum algebras.

DEFINITION 6.3. The **root space decomposition** of $U_v^{\pm}$ is the simultaneous eigenspace decomposition of $U_v^{\pm}$ with respect to the adjoint action:

$$Ad(t)u := tut^{-1} \quad (t = t_0, \ldots, t_{r-1}, v^d; u \in U_v^{\pm}).$$

That is,

$$U_{\pm\alpha} = \{ u \in U_v^{\pm} \mid Ad(t_i)u = v^{\pm\alpha(h_i)}u, Ad(v^d)u = v^{\pm\alpha(d)}u \}$$

for $\alpha \in Q_+ = \oplus_{i=1}^{r-1} \mathbb{Z}_{\geq 0}\alpha_i$.

The following proposition is obvious, and we omit the proof.

PROPOSITION 6.4. *We have* $U_v^+ = \oplus_{\alpha \in Q_+} U_\alpha$ *and* $U_v^- = \oplus_{\alpha \in Q_+} U_{-\alpha}$.

6.2. Integrable highest weight modules

Crystal bases of integrable modules are studied in detail in the case when the weights are "bounded from above". If such an integrable module is generated by a vector annihilated by e_i $(0 \leq i \leq r-1)$, then it is always irreducible, and it can be described as a factor module of U_v by an explicitly defined left ideal. In this section, we prove these facts.

These results are proved in Lusztig's book [**Lusztig**]. But in his book, Lusztig has introduced the quantum algebra using the radical of a symmetric bilinear form as part of the defining relations. Thus his definition is different from ours; however, we do not want to prove that his definition and ours coincide. Further, if we use the quantum version of the Casimir operator, the two results above may be proved using a proof similar to that in Kac's book [**K**] whose proof is for Kac-Moody Lie algebras. This is a more conceptual argument, but we want to bypass the definition of the quantum Casimir operator.

Our strategy is to assume the existence of the Casimir element for Kac-Moody algebras and to use a "specialization to $v = 1$" argument to deduce these two results. If we were to accept the existence of the quantum Casimir operator, then the proof would be shorter. For completeness, we also record the statement for the quantum Casimir operator in Theorem 6.8 below.

Before stating Theorem 6.8, we have to give some definitions.

DEFINITION 6.5. A weight λ is **dominant** if it belongs to

$$P_+ = \{\, \lambda \in \mathfrak{h}^* \mid \lambda(h_i) \in \mathbb{Z}_{\geq 0},\ \lambda(d) \in \mathbb{Z} \,\}.$$

The **fundamental weights** Λ_i $(0 \leq i < r)$ are defined by

$$\Lambda_i(h_j) = \delta_{ij}, \quad \Lambda_i(d) = 0.$$

Note that we have $P_+ = \oplus_{i=0}^{r-1}\mathbb{Z}_{\geq 0}\Lambda_i \oplus \mathbb{Z}\delta$, where $\delta = \sum_{i=0}^{r-1} \alpha_i$.

DEFINITION 6.6. Let M be a U_v-module. A **primitive vector** is a vector $m \in M_\lambda$ for some $\lambda \in P$ such that $e_i m = 0$ for $0 \leq i \leq r-1$. M is called a **highest weight module** if it is generated by a primitive vector as a U_v-module.

DEFINITION 6.7. For $\lambda \in P_+$, we define I_λ by

$$I_\lambda = \sum_{i=0}^{r-1} U_v e_i + U_v(v^d - v^{\lambda(d)}) + \sum_{i=0}^{r-1} U_v(t_i - v^{\lambda(h_i)}) + \sum_{i=0}^{r-1} U_v f_i^{1+\lambda(h_i)},$$

and set $V(\lambda) = U_v/I_\lambda$. If we denote $1 \pmod{I_\lambda} \in V(\lambda)$ by v_λ, then it is a primitive vector and generates $V(\lambda)$ as a U_v-module. We call v_λ the **highest weight vector** of $V(\lambda)$.

In the case of the enveloping algebra U of the Kac-Moody Lie algebra, we define $V(\lambda)$ as follows: Let

$$I_\lambda = \sum_{i=0}^{r-1} U e_i + U(d - \lambda(d)) + \sum_{i=0}^{r-1} U(h_i - \lambda(h_i)) + \sum_{i=0}^{r-1} U f_i^{1+\lambda(h_i)},$$

and $V(\lambda) = U/I_\lambda$.

In the theorem, we use a non-degenerate symmetric bilinear form $(\ ,\ )$ defined on $\mathfrak{h}^*$, and $|\lambda|^2$ stands for (λ, λ). If the Cartan matrix $A = (a_{ij})$ is of type $A_{r-1}^{(1)}$, then the bilinear form is given by

$$(\alpha_i, \alpha_j) = a_{ij}, \quad (\alpha_i, \Lambda_0) = \delta_{i0}, \quad (\Lambda_0, \Lambda_0) = 0.$$

We set $\rho = \sum_{i=0}^{r-1} \Lambda_i$. Now we state the theorem. See [**K**, Theorem 2.6] and [**Lusztig**, Proposition 6.1.7] for the Kac-Moody and quantum cases respectively.

THEOREM 6.8. *The following hold.*

(1) *Let M be an integrable highest weight U-module. Then there exists an operator C on M, called the* **Casimir operator**, *such that (i) C commutes with the U-action, and (ii) if $m \in M_\lambda$ is a primitive vector, then $Cm = (|\lambda+\rho|^2 - |\rho|^2)m$.*
(2) *Let M be an integrable highest weight U_v-module. Then there exists an operator C on M, called the* **quantum Casimir operator**, *such that (i) C commutes with the U_v-action, and (ii) if $m \in M_\lambda$ is a primitive vector, then $Cm = (v^{|\lambda+\rho|^2 - |\rho|^2})m$.*

LEMMA 6.9. *Suppose that $\lambda \in P_+$. Then the following hold for the U_v-modules $V(\lambda)$.*

(1) *Each $V(\lambda)$ is an integrable U_v-module.*
(2) *Each $V(\lambda)$ is irreducible.*

(3) *Let M be an integrable U_v-module and $m_\lambda \in M_\lambda$ a primitive vector. Then $\lambda \in P_+$ and $U_v m_\lambda \simeq V(\lambda)$. In particular, every integrable highest weight module is irreducible and $\{V(\lambda) \,|\, \lambda \in P_+\}$ is a complete set of integrable highest weight modules.*

PROOF. (1) As U_v has a weight space decomposition, it is obvious that $V(\lambda)$ also has a weight space decomposition. The multiplicity of a weight μ in $V(\lambda)$ is finite since

$$\dim V(\lambda)_\mu \leq |\{f_{i_1}\cdots f_{i_N} \mid \sum_{k=1}^{N} \alpha_{i_k} = \lambda - \mu\}|.$$

To show that $e_i^n m = 0, f_i^n m = 0$ $(n >> 0)$ for all m, we may assume that $m = f_{i_1}\cdots f_{i_N} \pmod{I_\lambda}$. That $e_i^n m = 0$ for $n >> 0$ is obvious since $V(\lambda)_\mu = 0$ for $\mu = \lambda - \sum \alpha_{i_k} + n\alpha_i$ with $n >> 0$. We prove $f_i^n m = 0$ for $n >> 0$ by induction on N. For $N = 0$, $m = v_\lambda$ and the assertion follows from $f_i^{1+\lambda(h_i)} m = 0$. Next we assume that the assertion holds for $m \equiv f_{i_1}\cdots f_{i_N}$. Then the assertion for $f_j m$ follows from the following formula.

$$f_i^n f_j = \begin{cases} [n] f_i f_j f_i^{n-1} - [n-1] f_j f_i^n & (i - j \equiv \pm 1) \\ f_j f_i^n & (\textit{otherwise}) \end{cases}$$

(2) Let R_1 be the localization of the ring $\mathbb{Q}[v, v^{-1}]$ with respect to the ideal $(v-1)$, and let W be a non-zero U_v-submodule of $V(\lambda)$. Since W has a weight space decomposition, we can choose a weight μ such that $W_\mu \neq 0$ and $W_{\mu+\alpha_i} = 0$ for all i. In particular, we can choose a non-zero vector $v_\mu \in V(\lambda)_\mu$ such that

$$e_i v_\mu = 0 \quad (0 \leq i \leq r-1).$$

We now remark that there exists an R_1-lattice $V_{R_1}(\lambda)$ of $V(\lambda)$ which is stable under the action of $e_i^{(n)}$ and $f_i^{(n)}$ $(0 \leq i < r, n \in \mathbb{N})$. To see this, let U_{R_1} be the R_1-subalgebra of U_v generated by $\{e_i^{(n)}, f_i^{(n)} | 0 \leq i < r, n \in \mathbb{N}\}$. Then $V_{R_1}(\lambda) := U_{R_1} v_\lambda$ is an R_1-lattice of $V(\lambda)$. We check that $e_i^{(m)} f_i^{(n)} - f_i^{(n)} e_i^{(m)}$ equals

$$\sum_{j=1}^{\min(m,n)} f_i^{(n-j)} \left(\prod_{k=1}^{j} \frac{v^{-m-n+2j-k+1} t_i - v^{m+n-2j+k-1} t_i^{-1}}{v^k - v^{-k}} \right) e_i^{(m-j)},$$

which implies that $V_{R_1}(\lambda)$ has a weight space decomposition $V_{R_1}(\lambda) = \oplus_\mu V_{R_1}(\lambda)_\mu$, and that the $V_{R_1}(\lambda)_\mu$ are finitely generated free R_1-modules.

Therefore, we may assume that $v_\lambda \not\equiv 0$ and $v_\mu \not\equiv 0 \pmod{(v-1)V_{R_1}(\lambda)}$ by multiplying by a suitable scalar in each case. We now specialize v to 1.

Then we have a representation of U on $V_{R_1}(\lambda)/(v-1)V_{R_1}(\lambda)$ with the following action. Note that the weight multiplicities are the same as in the quantum case.

$$e_i := e_i \pmod{(v-1)V_{R_1}(\lambda)}, \quad f_i := f_i \pmod{(v-1)V_{R_1}(\lambda)},$$

$$h_i := \frac{t_i - t_i^{-1}}{v - v^{-1}} \pmod{(v-1)V_{R_1}(\lambda)}, \; d := \frac{v^d - v^{-d}}{v - v^{-1}} \pmod{(v-1)V_{R_1}(\lambda)}.$$

This action ensures that $V_{R_1}(\lambda)/(v-1)V_{R_1}(\lambda)$ is integrable. Hence, if we let the Casimir operator act on $v_\lambda \pmod{(v-1)V_{R_1}(\lambda)}$ and on $v_\mu \pmod{(v-1)V_{R_1}(\lambda)}$, then we get $|\lambda+\rho|^2 = |\mu+\rho|^2$. (If we use Theorem 6.8(2), we obtain this equation directly from the existence of v_μ.) Because $V(\lambda)$ is integrable, v_μ is a primitive vector for a finite dimensional $U_v(\mathfrak{g}_i)$-submodule. In particular, we have $\mu(h_i) \in \mathbb{Z}_{\geq 0}$ for all i. Hence $\lambda, \mu \in P_+$.

We now note that $(\lambda - \mu, \lambda) \geq 0$, $(\lambda - \mu, \mu) \geq 0$ and $(\lambda - \mu, \rho) \geq 0$ satisfy

$$(\lambda - \mu, \lambda) + (\lambda - \mu, \mu) + 2(\lambda - \mu, \rho) = |\lambda + \rho|^2 - |\mu + \rho|^2 = 0.$$

Hence we have $(\lambda - \mu, \rho) = 0$, which implies that $\lambda = \mu$.

The above argument shows that a non-zero U_v-submodule of $V(\lambda)$ must coincide with $V(\lambda)$. Hence $V(\lambda)$ is irreducible.

(3) Since M is integrable, m_λ is a primitive vector of a finite dimensional $U_v(\mathfrak{g}_i)$-module. Hence we have $\lambda \in P_+$ and $f_i^{1+\lambda(h_i)} m_\lambda = 0$. As a result, we have a surjective U_v-homomorphism $V(\lambda) \to U_v m_\lambda \subset M$. By (2), we conclude that this is an isomorphism. □

6.3. Crystal bases of U_v^-

For simplicity, we assume that U_v is of type $A_{r-1}^{(1)}$ with $r \geq 3$ as before. We shall introduce the notion of the crystal basis for U_v^-. We follow the arguments in [**Kashiwara**]. The principal idea is to introduce an algebra called the Kashiwara algebra. Recall that $K = \mathbb{Q}(v)$.

DEFINITION 6.10. The **Kashiwara algebra** $\mathcal{B}_v$ is the K-algebra defined by generators e'_i and f_i, for $0 \leq i \leq r-1$, and the following relations.

$$e'_i f_j = v^{-\alpha_j(h_i)} f_j e'_i + \delta_{ij}$$

$$e'^2_i e'_j - (v + v^{-1}) e'_i e'_j e'_i + e'_j e'^2_i = 0 \quad (i - j \equiv \pm 1 \,(\mathrm{mod}\, r))$$
$$e'_i e'_j = e'_j e'_i \quad (otherwise)$$

$$f_i^2 f_j - (v + v^{-1}) f_i f_j f_i + f_j f_i^2 = 0 \quad (i - j \equiv \pm 1 \,(\mathrm{mod}\, r))$$
$$f_i f_j = f_j f_i \quad (otherwise)$$

To define Kashiwara operators on U_v^- we consider U_v^- as a $\mathcal{B}_v$-module. Then we may define the notion of a crystal basis on U_v^-. (This strategy also works for more general $\mathcal{B}_v$-modules.)

We start with the triangular decomposition of $\mathcal{B}_v$. The proof is similar to that of Theorem 6.2, so we leave it to the reader. Given this, Proposition 6.13 defines a $\mathcal{B}_v$-module structure on U_v^-.

DEFINITION 6.11. We denote by $\mathcal{B}_v^+$ and $\mathcal{B}_v^-$ the K-subalgebra of $\mathcal{B}_v$ generated by $\{e'_i\}_{0 \leq i \leq r-1}$ and the K-subalgebra of $\mathcal{B}_v$ generated by $\{f_i\}_{0 \leq i \leq r-1}$.

LEMMA 6.12. *We have the following.*

1. *The product map induces an isomorphism of vector spaces* $\mathcal{B}_v^- \otimes_K \mathcal{B}_v^+ \simeq \mathcal{B}_v$.
2. *Let* I'_+ *be the two-sided ideal of the tensor algebra* $T(\oplus K e'_i)$ *generated by*

$$\begin{cases} e'^2_i e'_j - (v + v^{-1}) e'_i e'_j e'_i + e'_j e'^2_i & (i - j \equiv \pm 1) \\ e'_i e'_j - e'_j e'_i & (otherwise) \end{cases},$$

and let I_- *be the two-sided ideal of the tensor algebra* $T(\oplus K f_i)$ *generated by*

$$\begin{cases} f_i^2 f_j - (v + v^{-1}) f_i f_j f_i + f_j f_i^2 & (i - j \equiv \pm 1) \\ f_i f_j - f_j f_i & (otherwise) \end{cases}.$$

Then we have the following isomorphisms of vector spaces. (The maps are obvious ones.)

$$T(\oplus_{i=0}^{r-1} K e_i')/I_+' \simeq \mathcal{B}_v^+, \quad T(\oplus_{i=0}^{r-1} K f_i)/I_- \simeq \mathcal{B}_v^-.$$

(3) *The map: $f_i \mapsto f_i$ induces an isomorphism of K-algebras between $\mathcal{B}_v^-$ and U_v^-.*

PROPOSITION 6.13. ([**Kashiwara**, Lemma 3.4.1, 3.4.2, 3.4.3])

(1) *For $P \in U_v^-$, there exist unique P', $P'' \in U_v^-$ which satisfy the following equation.*

$$[e_i,\, P] = \frac{t_i P' - t_i^{-1} P''}{v - v^{-1}}$$

We write $e_i'(P) = P''$.

(2) *Let e_i' be as in (1), and let f_i be the operator on U_v^- defined by the multiplication by f_i. Then U_v^- becomes a $\mathcal{B}_v$-module.*

(3) *As a $\mathcal{B}_v$-module, U_v^- has the following description as the factor module by a left ideal of $\mathcal{B}_v$.*

$$U_v^- \simeq \mathcal{B}_v / \sum_{i=0}^{r-1} \mathcal{B}_v e_i'$$

PROOF. (1) To show uniqueness we assume $\dfrac{t_i P_1' - t_i^{-1} P_1''}{v - v^{-1}} = \dfrac{t_i P_2' - t_i^{-1} P_2''}{v - v^{-1}}$. Then we have

$$t_i(P_1' - P_2') + t_i^{-1}(P_2'' - P_1'') = 0.$$

Applying Theorem 6.2, we get $P_1' = P_2', P_1'' = P_2''$. Hence uniqueness holds.

To show the existence of P' and P'', we consider the root space decomposition $U_v^- = \oplus U_{-\alpha}$. We now prove the existence of P' and P'' by induction on the height of roots $\mathrm{ht}(\alpha) = \sum m_i$ where $\alpha = \sum m_i \alpha_i$ ($m_i \in \mathbb{Z}_{\geq 0}$). If $P = 1$ then we may choose $P' = P'' = 0$. Assume the induction hypothesis for $P \in U_{-\alpha}$; there exist P', P'' which satisfy $[e_i,\, P] = \dfrac{t_i P' - t_i^{-1} P''}{v - v^{-1}}$. Then we have

$$\begin{aligned}[e_i,\, f_j P] &= [e_i,\, f_j] P + f_j [e_i,\, P] \\ &= \frac{t_i\left(\delta_{ij} P + v^{\alpha_j(h_i)} f_j P'\right) - t_i^{-1}\left(\delta_{ij} P + v^{-\alpha_j(h_i)} f_j P''\right)}{v - v^{-1}},\end{aligned}$$

and the induction can be continued. In particular, we have

$$e_i' f_j(P) = \delta_{ij} P + v^{-\alpha_j(h_i)} f_j e_i'(P). \tag{6.2}$$

(2) The commutation relation between e_i' and f_j follows from (6.2). The relations among the f_i follow from those for U_v^-. To show the relations for the e_i', we claim the following formula.

Assertion *Assume that $i - j \equiv \pm 1 \pmod r$, and set*

$$S = e_i'^2 e_j' - (v + v^{-1}) e_i' e_j' e_i' + e_j' e_i'^2.$$

Then $S f_k = v^{-\alpha_k(2h_i + h_j)} f_k S$ for all k.

Using the commutation relation for e'_i and f_j, we have

$$\begin{aligned}Sf_k = v^{-\alpha_k(2h_i+h_j)} f_k S + \delta_{ik}\left(v^{-\alpha_k(h_i+h_j)} + v^{-\alpha_k(h_j)} - v - v^{-1}\right) e'_i e'_j \\ + \delta_{jk}\left(1 + v^{-\alpha_k(2h_i)} - v^{1-\alpha_k(h_i)} - v^{-1-\alpha_k(h_i)}\right) e'^2_i \\ + \delta_{ik}\left(1 + v^{-\alpha_k(h_i)} - v^{1-\alpha_k(h_i+h_j)} - v^{-1-\alpha_k(h_i+h_j)}\right) e'_j e'_i.\end{aligned}$$

Hence, the assertion is obvious when $k \neq i, j$. If $k = i$ or $k = j$, it is enough to observe that all of the terms other than the first are 0.

Using this formula and the fact that $S(1) = 0$, we see that $S(f_{i_1} \cdots f_{i_N}) = 0$ by induction on N.

Next we assume that $i - j \not\equiv \pm 1$. Since $e_i e_j = e_j e_i$, we have $e'_i e'_j = e'_j e'_i$ by comparing the coefficients of $t_i^{-1} t_j^{-1}$ in both sides of $[e_i, [e_j, P]] = [e_j, [e_i, P]]$. We have proved that U_v^- becomes a $\mathcal{B}_v$-module via the given operators.

(3) Let $\mathcal{B}_v^-$ and $\mathcal{B}_v^+$ be as in Lemma 6.12. Then Lemma 6.12(1) says that $\mathcal{B}_v \simeq \mathcal{B}_v^- \otimes \mathcal{B}_v^+$. Thus, the map $\mathcal{B}_v \to U_v^-$ defined by $a \mapsto a \cdot 1$ induces a surjection

$$\mathcal{B}_v / \sum \mathcal{B}_v e'_i \to U_v^-. \tag{6.3}$$

As a vector space, the left hand side is isomorphic to $\mathcal{B}_v^-$ and the surjection induces a surjection $\mathcal{B}_v^- \to U_v^- : f_i \mapsto f_i$. If we compose this map with the isomorphism of algebras $U_v^- \simeq \mathcal{B}_v^-$ given by Lemma 6.12(3), then we obtain the identity map on $\mathcal{B}_v^-$. Hence, we conclude that U_v^- has the desired description as the quotient of $\mathcal{B}_v$ by a left ideal. □

PROPOSITION 6.14 ([**Kashiwara**, Proposition 3.2.1]). *Let P_i be an operator on U_v^- given by*

$$P_i = \sum_{n=0}^{\infty} (-1)^n v^{-n(n-1)/2} f_i^{(n)} e'^n_i.$$

The operator P_i is well-defined and the following hold.

(1) *Assume that u is an element of U_v^-. If $u = \sum_{n\geq 0} f_i^{(n)} u_n$ ($u_n \in \operatorname{Ker} e'_i$) then $u_n = v^{n(n-1)/2} P_i(e'^n_i u)$. Conversely, if we set $u_n = v^{n(n-1)/2} P_i(e'^n_i u)$ then $u_n \in \operatorname{Ker} e'_i$ and we have $u = \sum_{n\geq 0} f_i^{(n)} u_n$.*

(2) $U_v^- = \bigoplus_{n\geq 0} f_i^{(n)} \operatorname{Ker} e'_i$.

PROOF. (1) We prove a set of formulas. In the following, we use the notation $\begin{bmatrix} n \\ k \end{bmatrix} = \dfrac{[n]!}{[k]![n-k]!}$.

Assertion *The following formulas hold.*

(i) $e'_i f_i^{(n)} = v^{-2n} f_i^{(n)} e'_i + v^{-n+1} f_i^{(n-1)}$.

(ii) $e'^n_i f_i = v^{-2n} f_i e'^n_i + v^{-n+1} [n] e'^{n-1}_i$.

(iii) $e'_i P_i = 0,\ P_i f_i = 0$.

(iv) $\displaystyle\sum_{n=0}^{\infty} v^{n(n-1)/2} f_i^{(n)} P_i e'^n_i = 1$.

(v) $e_i'^m f_i^{(n)} = \sum_{l=0}^{\min(m,n)} v^{-2mn+(m+n)l-l(l-1)/2} \begin{bmatrix} m \\ l \end{bmatrix} f_i^{(n-l)} e_i'^{m-l}$.

Parts (i) and (ii) are proved by induction on n.
(iii) To show that $e_i'P_i = 0$, we replace $e_i'f_i^{(n)}$ by $v^{-2n}f_i^{(n)}e_i' + v^{-n+1}f_i^{(n-1)}$ in

$$e_i'P_i = \sum (-1)^n v^{-n(n-1)/2} e_i' f_i^{(n)} e_i'^n.$$

If we use (ii) instead of (i), we obtain $P_i f_i = 0$.
(iv) Expanding the sum in (iv) using the definition of P_i we obtain

$$\sum_{N=0}^{\infty} v^{N(N-1)/2} \left(\sum_{n=0}^{N} (-1)^n v^{(1-N)n} \begin{bmatrix} N \\ n \end{bmatrix} \right) f_i^{(N)} e_i'^N.$$

Now use the following formula, which may be proved by induction on N:

$$\sum_{n=0}^{N} (-1)^n \begin{bmatrix} N \\ n \end{bmatrix} x^n = \prod_{n=1}^{N} (1 - v^{2n-N-1}x).$$

As a result, we know that all of the terms with $N \geq 1$ vanish; this gives (iv).
(v) This is proved by induction on m. The formulas needed in the proof are (ii) and

$$v^l \begin{bmatrix} m \\ l \end{bmatrix} + v^{l-m-1} \begin{bmatrix} m \\ l-1 \end{bmatrix} = \begin{bmatrix} m+1 \\ l \end{bmatrix}.$$

We now prove (1). If we set $u_n = v^{n(n-1)/2} P_i(e_i'^n u)$, then we have

$$u = \sum f_i^{(n)} u_n \quad (u_n \in \operatorname{Ker} e_i')$$

by (iii) and (iv). To show the uniqueness of this expression, we assume that we are given an expression $u = \sum f_i^{(n)} u_n$ ($u_n \in \operatorname{Ker} e_i'$), where $u_n = 0$ ($n >> 0$). Using (v), $P_i(e_i'^m u) = \sum_{n=0}^{\infty} P_i e_i'^m f_i^{(n)} u_n$ may be rewritten as follows.

$$P_i(e_i'^m u) = \sum_{n \geq 0} \sum_{l=0}^{\min(m,n)} v^{-2mn+(m+n)l-l(l-1)/2} \begin{bmatrix} m \\ l \end{bmatrix} P_i f_i^{(n-l)} e_i'^{m-l} u_n$$

Since $u_n \in \operatorname{Ker} e_i'$, all of the terms other than $l = m \leq n$ vanish. Hence we have

$$P_i(e_i'^m u) = \sum_{n \geq m} v^{-mn+m(m+1)/2} P_i f_i^{(n-m)} u_n.$$

Since $P_i f_i^{(n-m)} = 0$ ($n > m$) by (iii), the right hand side is $v^{-m(m-1)/2} P_i u_m$. By definition, $P_i u_m = \sum_{n \geq 0} (-1)^n v^{-n(n-1)/2} f_i^{(n)} e_i'^n u_m = u_m$ so we conclude that $u_m = v^{m(m-1)/2} P_i(e_i'^m u)$. Hence the uniqueness of the expression is proved.
(2) This follows from (1). □

We can now define the Kashiwara operators and the crystal basis of U_v^-.

DEFINITION 6.15. We write $u = \sum_{n\geq 0} f_i^{(n)} u_n$ ($u_n \in \operatorname{Ker} e_i'$) as above. Define operators $\tilde{e}_i$ and $\tilde{f}_i$ on U_v^- by

$$\tilde{e}_i u = \sum f_i^{(n-1)} u_n, \quad \tilde{f}_i u = \sum f_i^{(n+1)} u_n.$$

These operators are the **Kashiwara operators** on U_v^-.

DEFINITION 6.16. A **crystal basis** of U_v^- is a pair (L, B) which satisfies the following conditions.

(1) L is a full rank R-lattice of U_v^-, and $\tilde{e}_i L \subset L$, $\tilde{f}_i L \subset L$ for all i.
(2) B is a $\mathbb{Q}$-basis of L/vL, and $\tilde{e}_i B \subset B \cup \{0\}$, $\tilde{f}_i B \subset B$ for all i.
(3) Set $L_{-\alpha} = U_{-\alpha} \cap L$, $B_{-\alpha} = (L_{-\alpha}/vL_{-\alpha}) \cap B$. Then both L and B have the root space decompositions
$$L = \bigoplus_{\alpha \in Q_+} L_{-\alpha}, \quad B = \bigsqcup_{\alpha \in Q_+} B_{-\alpha}.$$
(4) For $b \in B$, $\tilde{e}_i b \in B$ implies $b = \tilde{f}_i \tilde{e}_i b$ for all i.

CHAPTER 7

The canonical basis

7.1. A review of Lusztig's canonical basis

We have proved several properties of the crystal basis in the previous chapters, but we have not said anything about its existence yet. Let us consider the crystal basis of U_v^- first. In this case, there exists the "canonical basis" constructed by Lusztig and this leads to the existence of the crystal basis of U_v^-. In the case of integrable U_v-modules, we may also obtain the crystal basis from the canonical basis, which may be identified with a subset of the canonical basis of U_v^-.

This dicovery of the relationship between the crystal basis and the canonical basis was a big event. (See [**Lusztig**, Notes on Part II] for his historical comments.)

In this chapter, we start with two theorems which state basic properties of the canonical basis and we deduce several consequences of these theorems. The existence of the crystal bases will be proved in the subsequent two chapters.

We do not provide the proofs of these two theorems since they require sophisticated geometric machinery. If a reader is unhappy about the absence of the proofs, we recommend looking at the third section of Chapter 14. In this section, the geometric construction of the Hall algebra of a cyclic quiver is explained; this is a good source for the ideas of the construction of the quantum algebra and its canonical basis.

In the following, we set $A = \mathbb{Z}[v, v^{-1}]$. We begin with some definitions.

DEFINITION 7.1. The **Kostant-Lusztig form** of the quantum algebra U_v^- is the A-subalgebra of U_v^- generated by $f_i^{(n)}$ $(n = 0, 1, \dots)$. We denote this algebra by U_A^-.

If we set $(U_A^-)_{-\alpha} := U_A^- \cap U_{-\alpha}$, then we have

$$U_A^- = \bigoplus_{\alpha \in Q_+} (U_A^-)_{-\alpha}.$$

In other words, U_A^- has a root space decomposition.

DEFINITION 7.2. The **bar involution** $P \mapsto \bar{P}$ on U_v^- is the $\mathbb{Q}$-algebra automorphism of U defined by the following properties.

$$\overline{PQ} = \bar{P}\bar{Q}, \quad \bar{f}_i = f_i, \quad \bar{v} = v^{-1}.$$

We summarize the properties of Lusztig's canonical basis needed in these lectures in Theorem 7.3 and Theorem 7.5 below.

Lusztig has constructed U_v^- in a geometric way [**Lusztig**, Theorem 13.2.11], and has defined the canonical basis of U_v^- using perverse sheaves, which naturally appear in this construction [**Lusztig**, 14.4.1, 14.4.4, 14.4.6].

By translating the results so obtained into algebraic language, we have the properties of the canonical basis stated in these theorems.

Before stating the theorems, I give pointers to the proofs of these properties. The references are all in [**Lusztig**].

The pointers for Theorem 7.3 are:
(1) [Theorem 14.4.3(d)(e), 14.4.4],
(2) [Proposition 12.5.2(b), 12.6.4, 13.1.11(a)],
(3) [9.3.1, Proposition 10.3.2, Remark 10.3.3, Theorem 14.3.2(e)],
(4) [9.3.1, 13.1.2(a)(associativity of $ind^V_{T,W}$)],
(5) [9.3.1(b), Lemma 12.5.1(c)],
(6) [10.3.4, Theorem 14.4.3(c), 14.4.4].

The pointers for Theorem 7.5 are:
(1) [13.1.6], (2) [Lemma 12.2.2, 13.1.3, 13.1.7],
(3) [Lemma 12.3.6, 13.1.12(d)], (4) [Lemma 12.5.3, 13.1.11(b)(c)].

As in the previous chapters, our restriction that the Cartan matrix has type $A^{(1)}_{r-1}$ is still not essential. In fact these theorems are valid whenever the Cartan matrix is symmetric. If we modify the statements appropriately, these theorems are valid for more general symmetrizable Cartan matrices.

THEOREM 7.3. *There exists an A-free basis B of U^-_A with the following properties.*

(1) *Set $B_{-\alpha} := U_{-\alpha} \cap B$. Then $B_{-\alpha}$ is an A-free basis of $(U^-_A)_{-\alpha}$ and*

$$B_0 = \{1\}, \quad B = \bigsqcup_{\alpha \in Q_+} B_{-\alpha}.$$

(2) *Each basis element of B is bar invariant; that is, $\bar{b} = b$ ($b \in B$).*
(3) *There exists a partition $B = \sqcup_{n \geq 0} B_{i,n}$ and a bijection $\pi_n : B_{i,0} \simeq B_{i,n}$ which induces a bijection $B_{i,0} \cap B_{-\alpha} \simeq B_{i,n} \cap B_{-\alpha - n\alpha_i}$ such that*

$$f_i^{(n)} b - \pi_n(b) \in \sum_{\substack{b' \in B_{i,m} \\ m > n}} Ab' \quad (b \in B_{i,0}).$$

(4) *We have the inclusion*

$$\sum_{\substack{b \in B_{i,m} \\ m \geq n}} Ab \subset \bigoplus_{\alpha \in n\alpha_i + Q_+} (U^-_A)_{-\alpha},$$

and $\sum_{\substack{b \in B_{i,m} \\ m \geq n}} Ab$ *is stable under left multiplication by $f_i^{(k)}$ ($k \in \mathbf{N}$).*

(5) *For each $b \in B \setminus \{1\}$, there exists an index i such that $b \notin B_{i,0}$.*
(6) *Let σ be the anti-involution of U^-_v defined by $\sigma(f_i) = f_i$. Then $\sigma(B) = B$.*

DEFINITION 7.4. The **canonical basis** is the A-free basis B of the quantum algebra U^-_A introduced in Theorem 7.3.

THEOREM 7.5. *Let $\Delta_+ : U^-_v \to U^-_v \otimes U^-_v$ be the algebra homomorphism defined by*

$$\Delta_+(f_i) = 1 \otimes f_i + f_i \otimes t_i^{-1}.$$

We set $t_\alpha := t_0^{m_0} \cdots t_{r-1}^{m_{r-1}}$ if $\alpha = \sum m_i \alpha_i$, and define $\Delta_{\alpha,\beta} : U^-_v \to U_{-\alpha} \otimes U_{-\beta}$ by

$$\Delta_+(u) = \sum_{\alpha,\beta \in Q_+} (1 \otimes t_{-\alpha}) \Delta_{\alpha,\beta}(u) \quad (\Delta_{\alpha,\beta}(u) \in U_{-\alpha} \otimes U_{-\beta}).$$

Then there exists a non-singular symmetric bilinear form $(\ ,\ )$ *on* U_v^- *with the following properties.*

(1) $(U_{-\alpha}, U_{-\beta}) = 0\ (\alpha \neq \beta)$, $(U_{-\alpha}, U_{-\alpha}) \subset \mathbb{Q}((v))$.

(2) *If* $P \in U_{-\alpha}$ *and* $Q \in U_{-\beta}$ *then*

$$(PQ, R) = (P \otimes Q, \Delta_{\alpha,\beta}(R))\quad (R \in U_v^-).$$

(3) $(1, 1) = 1$, $(f_i, f_i) = \dfrac{1}{1-v^2}$.

(4) *For* $b, b' \in B$, *we have* $(b, b') \in \mathbb{Z}[[v]]$ *and* $(b, b') \equiv \delta_{bb'} \pmod{v\mathbb{Z}[[v]]}$.

We prove several consequences of the theorems.

LEMMA 7.6. *The canonical basis has the following properties.*

(1) *The* A*-module* $\sum_{m\geq n} f_i^{(m)} U_A^-$ *has* $\sqcup_{m\geq n} B_{i,m}$ *as an* A*-free basis.*

(2) *For each basis element* $b \in B\backslash\{1\}$, *there exists an index* i *such that* $b \in f_i U_v^-$.

PROOF. (1) We prove this in the following steps [**Lusztig**, Theorem 14.3.2]. As before, the height of a root $\alpha = \sum m_i\alpha_i \in Q_+$ is $\mathrm{ht}(\alpha) = \sum m_i$.

(step 1) *The following inclusion holds.*

$$\sum_{m\geq n} f_i^{(m)}\left(U_A^-\right)_{-\alpha+m\alpha_i} \supset \sum_{\substack{b\in B_{i,m}\cap B_{-\alpha}\\ m\geq n}} Ab.$$

This is proved by downward induction on n. Since Theorem 7.3(4) implies $B_{i,m} \subset \oplus_{\alpha'\in m\alpha_i+Q_+} U_{-\alpha'}$, $B_{i,m}\cap B_{-\alpha}$ is empty if $\alpha - m\alpha_i \notin Q_+$. Hence the above inclusion holds for sufficiently large n.

Assume that the inclusion is already proved for $n+1, n+2, \ldots$. Take an element $b \in B_{i,m}\cap B_{-\alpha}$ where $m \geq n$. By Theorem 7.3(3) we have

$$b \in f_i^{(m)}\pi_m^{-1}(b) + \sum_{b'\in B_{i,l}\cap B_{-\alpha}, l>m} Ab',$$

and, by the induction hypothesis,

$$\sum_{b'\in B_{i,l}\cap B_{-\alpha}, l>m} Ab' \subset \left(\sum_{l>m} f_i^{(l)} U_A^-\right)_{-\alpha}.$$

Recalling that $f_i^{(m)}\pi_m^{-1}(b) \in f_i^{(m)}\left(U_A^-\right)_{-\alpha+m\alpha_i}$, we have the inclusion for n.

(step 2) *The following inclusion holds.*

$$\sum_{m\geq n} f_i^{(m)}\left(U_A^-\right)_{-\alpha+m\alpha_i} \subset \sum_{\substack{b\in B_{i,m}\cap B_{-\alpha}\\ m\geq n}} Ab.$$

This is proved by induction on $\mathrm{ht}(\alpha)$. First assume that $\mathrm{ht}(\alpha) = 0$. Then the left hand side is 0 if $n > 0$, and A if $n = 0$. Hence Theorem 7.3(1) gives the result.

Next assume that the inclusion holds for all α' with $\mathrm{ht}(\alpha') < \mathrm{ht}(\alpha)$. Since $B_{-\alpha+m\alpha_i}$ is a basis of $(U_A^-)_{-\alpha+m\alpha_i}$, it suffices to show that if $b \in B_{-\alpha+m\alpha_i}$ and

$m \geq n$, then $f_i^{(m)} b$ belongs to the right hand side A-module. Assume that $b \in B_{i,t}$. If $t = 0$, then Theorem 7.3(3) says that

$$f_i^{(m)} b \in \pi_m(b) + \sum_{b' \in B_{i,k} \cap B_{-\alpha}, k > m} A b'.$$

Hence, $\pi_m(b) \in B_{i,m} \cap B_{-\alpha}$ and $m \geq n$ imply the result.

Next assume that $t > 0$. Then (step 1) says that

$$b \in \sum_{k \geq t} f_i^{(k)} (U_A^-)_{-\alpha + (k+m)\alpha_i}.$$

Since $k \geq t (> 0)$, we may apply the induction hypothesis to $\alpha - k\alpha_i$ and we have

$$f_i^{(m)} (U_A^-)_{-\alpha + (k+m)\alpha_i} \subset \sum_{b' \in B_{i,l} \cap B_{-\alpha + k\alpha_i}, l \geq n} A b'.$$

Since $f_i^{(m)} b \in \sum_{k \geq t} f_i^{(k)} f_i^{(m)} (U_A^-)_{-\alpha + (k+m)\alpha_i}$, we conclude that

$$f_i^{(m)} b \in \sum_{k \geq t} f_i^{(k)} \left(\sum_{b' \in B_{i,l} \cap B_{-\alpha + k\alpha_i}, l \geq n} A b' \right),$$

where the right hand side is contained in $\sum_{b' \in B_{i,l} \cap B_{-\alpha}, l \geq n} A b'$ by Theorem 7.3(4).
(2) By Theorem 7.3(5), we may choose i such that $b \in \sum_{b' \in B_{i,m}, m > 0} A b'$. Hence (1) implies that

$$b \in \sum_{m \geq 1} f_i^{(m)} U_A^-,$$

and the result follows. □

LEMMA 7.7. *(1)* $(f_i P, Q) = \dfrac{1}{1 - v^2} (P, e_i' Q)$.
(2) If $u \in \bigoplus_{b \in B} \mathbb{Z}[v] b$, *then we have* $(u, u) \in \mathbb{Z}_{\geq 0} + v\mathbb{Z}[[v]]$.

PROOF. (1) Let $P \in U_{-\alpha}$ and $Q \in U_{-\alpha - \alpha_i}$. Then, by Theorem 7.5(1)(2), we have

$$(f_i P, Q) = (f_i \otimes P, \Delta_{\alpha_i, \alpha}(Q)).$$

We define $d_i(Q)$ by $f_i \otimes d_i(Q) = \sum_\alpha \Delta_{\alpha_i, \alpha}(Q)$. In other words, we pick up the terms of the form $f_i \otimes -$ appearing in $\Delta_+(Q)$ and write

$$\Delta_+(Q) = f_i \otimes t_i^{-1} d_i(Q) + \cdots.$$

Then, we have

$$(f_i P, Q) = (f_i, f_i)(P, d_i(Q)) = \frac{1}{1 - v^2} (P, d_i(Q)).$$

To show that $d_i(Q)$ coincides with $e_i' Q$, we consider

$$\Delta_+(f_j Q) = (1 \otimes f_j + f_j \otimes t_j^{-1}) \left(\sum_{\alpha, \beta} (1 \otimes t_\alpha^{-1}) \Delta_{\alpha, \beta}(Q) \right),$$

and pick up the terms of the form $f_i \otimes -$. Then we have

$$f_i \otimes t_i^{-1} d_i(f_j Q) = f_i \otimes f_j t_i^{-1} d_i(Q) + \delta_{ij} \left(f_i \otimes t_i^{-1} \right) \left(\sum_\beta \Delta_{0,\beta}(Q) \right).$$

Using $\sum_\beta \Delta_{0,\beta}(Q) = 1 \otimes Q$, we obtain

$$d_i(f_j Q) = v^{-\alpha_j(h_i)} f_j d_i(Q) + \delta_{ij} Q.$$

Now we compare this with $e'_i f_j = v^{-\alpha_j(h_i)} f_j e'_i + \delta_{ij}$. Then we may prove $d_i(Q) = e'_i Q$ inductively, starting with $d_i(1) = e'_i(1) = 0$.
(2) This is obvious by Theorem 7.5(4). □

7.2. The lattice of the canonical basis

Recall that $R = \mathbb{Q}[v]_{(v)}, K = \mathbb{Q}(v)$. By expanding elements of R and K into formal power series in v, we regard R and K as subsets of $\mathbb{Q}[[v]]$ and $\mathbb{Q}((v))$ respectively.

In this section, we prove results about the R-lattice generated by the canonical basis. We need these results in the next chapter to prove that we may obtain the crystal basis of U_v^- from the canonical basis.

LEMMA 7.8. *Set $L = \oplus_{b \in B} Rb$ and $L_A = \oplus_{b \in B} \mathbb{Z}[v] b$. Then the following hold.*

(1) *U_A^- is stable under the action of e'_i and $f_i^{(n)}$.*
(2) $U_A^- = \bigoplus_{n \geq 0} f_i^{(n)} \left(U_A^- \cap \operatorname{Ker} e'_i \right)$.
(3) $(f_i^{(n)} u, f_i^{(m)} u') = \delta_{mn} \dfrac{(u, u')}{(1 - v^{2n}) \cdots (1 - v^2)}$ *if $u, u' \in \operatorname{Ker} e'_i$.*
(4) $L_A = \left\{ u \in U_A^- \mid (u, u) \in \mathbb{Q}[[v]] \right\}$.
(4') $L = \{ u \in U_v^- \mid (u, u) \in \mathbb{Q}[[v]] \}$.
(5) $vL_A = \left\{ u \in U_A^- \mid (u, u) \in v\mathbb{Q}[[v]] \right\}$.
(5') $vL = \{ u \in U_v^- \mid (u, u) \in v\mathbb{Q}[[v]] \}$.
(6) $L_A = \bigoplus_{n \geq 0} f_i^{(n)} (L_A \cap \operatorname{Ker} e'_i)$.
(6') $L = \bigoplus_{n \geq 0} f_i^{(n)} (L \cap \operatorname{Ker} e'_i)$.

PROOF. (1) It is obvious that U_A^- is stable under the action of $f_i^{(n)}$. To show that U_A^- is stable under the action of e'_i, it is enough to prove $e'_i f_{i_1}^{(n_1)} \cdots f_{i_N}^{(n_N)} \in U_A^-$ by induction on N. This follows from the formula

$$e'_i f_j^{(n)} = v^{-n\alpha_j(h_i)} f_j^{(n)} e'_i + \delta_{ij} v^{-n+1} f_j^{(n-1)}.$$

(If $i = j$, this is nothing but Assertion(i) in the proof of Proposition 6.14(1). If $i \neq j$, it is obvious since $e'_i f_j = v^{-\alpha_j(h_i)} f_j e'_i$.)
(2) In Proposition 6.14(1), we showed that we may express $u \in U_v^-$ as

$$u = \sum_{n \geq 0} f_i^{(n)} u_n \quad (u_n \in \operatorname{Ker} e'_i)$$

in a unique way, where u_n is given by $u_n = v^{n(n-1)/2} P_i({e'_i}^n u)$ and

$$P_i = \sum_{n=0}^{\infty} (-1)^n v^{-n(n-1)/2} f_i^{(n)} {e'_i}^n.$$

Since $P_i e_i'^n U_A^- \subset U_A^-$ by (1), $u_n \in U_A^- \cap \operatorname{Ker} e_i'$ whenever $u \in U_A^-$. Thus we have

$$U_A^- \subset \bigoplus_{n\geq 0} f_i^{(n)} \left(U_A^- \cap \operatorname{Ker} e_i' \right).$$

The opposite inclusion is obvious.
(3) This is obvious when $n = m = 0$. If $n > 0, m = 0$, Lemma 7.7(1) implies that

$$(f_i^{(n)} u,\, u') = \frac{1}{[n]} \frac{1}{1-v^2} (f_i^{(n-1)} u,\, e_i' u') = 0.$$

The case $n = 0, m > 0$ follows by symmetry. We assume that $n > 0$ and $m > 0$. The proof is by induction on $m + n$. Since Lemma 7.7(1) implies

$$(f_i^{(n)} u,\, f_i^{(m)} u') = \frac{1}{[n]} \frac{1}{1-v^2} (f_i^{(n-1)} u,\, e_i' f_i^{(m)} u'),$$

and $e_i' f_i^{(m)} = v^{-2m} f_i^{(m)} e_i' + v^{-m+1} f_i^{(m-1)}$ (Proposition 6.14 Assertion(i)), the induction hypothesis proves the result.
(4)(4') The proof is almost the same for both cases. We first prove (4'). By Theorem 7.5(4) we have

$$L = \bigoplus_{b\in B} Rb \subset \{ u \in U_v^- \mid (u, u) \in \mathbb{Q}[[v]] \}.$$

To show the opposite inclusion, we express $u \in U_v^-$ as

$$u = \sum_{b\in B} c_b(v) b \quad (c_b(v) \in K),$$

and define r_b and l_b by

$$c_b(v) - r_b v^{l_b} \in v^{l_b+1} R \quad (r_b \in \mathbb{Q}^\times, l_b \in \mathbb{Z}).$$

If $c_b(v) = 0$, we set $r_b = 1$ and $l_b = \infty$. Let $l = \min\{l_b\}_{b\in B}$. Then $(b,\, b') \in \delta_{bb'} + v\mathbb{Q}[[v]]$ implies that

$$(u,\, u) = \left(\sum_{l_b = l} r_b^2 \right) v^{2l} + (\textit{higher degree terms}) .$$

Thus $\sum_{l_b=l} r_b^2 > 0$ and $(u,\, u) \in \mathbb{Q}[[v]]$ imply $l \geq 0$, and $c_b(v) \in R$ as required.

To prove (4), we follow the same argument, but we may assume $c_b(v) \in A$ since $u \in U_A^-$. Hence we have $l \geq 0$ and $c_b(v) \in \mathbb{Z}[v]$.
(5)(5') The proof is the same as that of (4)(4').
(6)(6') Since the proofs are similar, we only prove (6'). Take $u \in L$. Then we have $(u,\, u) \in \mathbb{Q}[[v]]$ by (4'). We express u as $u = \sum f_i^{(n)} u_n$ ($u_n \in \operatorname{Ker} e_i'$) as in Proposition 6.14. We describe u_n as a linear combination of elements of the canonical basis B. Expanding the coeffients of u_n with respect to B into Laurent series in v, we get an expansion of u_n into Laurent series in v. If we denote its lowest degree term by $x_n v^{l_n}$ ($x_n \in (\oplus_{b\in B} \mathbb{Q}b) \setminus \{0\}$), then we have

$$u_n = x_n v^{l_n} + (\textit{higher degree terms}).$$

Let $l = \min\{l_n\}_{n\geq 0}$ and let the coefficient of v^{2l} in $(x_n,\, x_n) v^{2l_n}$ be r_n. Then $r_n \geq 0$, with strict inequality for those n which satisfy $l_n = l$, and

$$(u_n,\, u_n) - r_n v^{2l} \in v^{2l+1} \mathbb{Q}[[v]].$$

Since Lemma 7.8(3) implies that

$$(u, u) = \sum_{n \geq 0} \frac{(u_n, u_n)}{(1 - v^{2n}) \cdots (1 - v^2)},$$

we have

$$(u, u) - \left(\sum_{n \geq 0} r_n \right) v^{2l} \in v^{2l+1} \mathbb{Q}[[v]], \quad \sum_{n \geq 0} r_n > 0.$$

Since $(u, u) \in \mathbb{Q}[[v]]$ implies $l \geq 0$, we have $l_n \geq 0$ and $(u_n, u_n) \in \mathbb{Q}[[v]]$. Now we apply (4') to conclude that $u_n \in L$. We have proved

$$L \subset \bigoplus_{n \geq 0} f_i^{(n)} (L \cap \operatorname{Ker} e_i') .$$

To prove the opposite inclusion, we take $u \in L \cap \operatorname{Ker} e_i'$. Then we have $(u, u) \in \mathbb{Q}[[v]]$ by (4'), and (3) implies that

$$(f_i^{(n)} u, f_i^{(n)} u) = \frac{(u, u)}{(1 - v^{2n}) \cdots (1 - v^2)} \in \mathbb{Q}[[v]].$$

Using (4') again, we may conclude that $f_i^{(n)} u \in L$. □

CHAPTER 8

Existence and uniqueness (part I)

8.1. Preparatory lemmas

In this chapter, we shall prove the existence and uniqueness of the crystal basis of U_v^-. The following lemma shows that the canonical basis gives an R-free basis of a crystal lattice. (See [**Lusztig**, (1) Lemma 17.3.4, (2) Proposition 18.1.7].)

LEMMA 8.1. *Let B be the canonical basis and $L = \oplus_{b\in B} Rb$ as before. Define $L(\infty)$ by*

$$L(\infty) = \sum R\tilde{f}_{i_1} \cdots \tilde{f}_{i_N} 1,$$

where the sum is over all $N \in \mathbb{Z}_{\geq 0}$ and all $(i_1, \dots, i_N) \in (\mathbb{Z}/r\mathbb{Z})^N$. Then we have the following.

(1) *L is stable under the action of $\tilde{e}_i$ and $\tilde{f}_i$.*
(2) *We have $L(\infty) = L$. In particular, the canonical basis is an R-free basis of $L(\infty)$.*

PROOF. (1) By Lemma 7.8(6') we have $L = \oplus_{n\geq 0} f_i{}^{(n)} (L \cap \operatorname{Ker} e_i')$. Hence this is obvious.
(2) Set $L_{-\alpha} = L \cap U_{-\alpha}$. Then we have $L = \oplus_{\alpha\in Q_+} L_{-\alpha}$ by Theorem 7.3(1). Next we set $L(\infty)_{-\alpha} = L(\infty) \cap U_{-\alpha}$. Then $L(\infty) = \oplus_{\alpha\in Q_+} L(\infty)_{-\alpha}$ by definition. Thus it suffices to prove that $L_{-\alpha} = L(\infty)_{-\alpha}$ by induction on $\operatorname{ht}(\alpha)$.

Since $1 \in B$, the claim holds if $\operatorname{ht}(\alpha) = 0$. Assume that $1 \neq b \in B_{-\alpha}$. Then, by Lemma 7.6(2), there exists an index i such that $b \in f_i U_v^-$. We write b as $b = \sum_{n\geq 0} f_i^{(n)} u_n$ ($u_n \in \operatorname{Ker} e_i'$).

(step 1) We have $u_0 \in vL_{-\alpha}$.

Since $u_0 = b - \sum_{n>0} f_i^{(n)} u_n \in f_i U_v^-$, we may write $u_0 = f_i u_0'$ and we have

$$(u_0, u_0) = (f_i u_0', u_0) = \frac{1}{1 - v^2} (u_0', e_i' u_0) = 0.$$

Hence we have $u_0 \in vL_{-\alpha}$ by Lemma 7.8(5').

(step 2) *There exists an integer $n_0 \geq 1$ such that*

$$f_i^{(n)} u_n \in vL_{-\alpha} \ (n \neq n_0), \quad f_i^{(n_0)} u_{n_0} \in L(\infty)_{-\alpha}.$$

We have $u_n \in L_A = \sum_{b\in B} \mathbb{Z}[v] b$ by Lemma 7.8(6), and

$$(b, b) \equiv \sum_{n\geq 1} (u_n, u_n) \pmod{v\mathbb{Z}[[v]]}$$

by Lemma 7.8(3) and (step 1). Recall that $(u_n, u_n) \pmod{v\mathbb{Z}[[v]]} \in \mathbb{Z}_{\geq 0}$ (Lemma 7.7(2)) and $(b, b) \equiv 1 \pmod{v\mathbb{Z}[[v]]}$. So we may choose $n_0 \geq 1$ such that we have

$$(u_{n_0}, u_{n_0}) \equiv 1, \ (u_n, u_n) \equiv 0 \ (n \neq n_0).$$

We use Lemma 7.8(3) again and conclude that $(f_i^{(n)} u_n, f_i^{(n)} u_n) \equiv 0 \ (n \neq n_0)$. Using Lemma 7.8(4')(5'), we obtain

$$u_{n_0} \in L_{-\alpha+n_0\alpha_i}, \quad f_i^{(n)} u_n \in vL_{-\alpha} \ (n \neq n_0).$$

Thus the induction hypothesis $L_{-\alpha+n_0\alpha_i} = L(\infty)_{-\alpha+n_0\alpha_i}$ implies that

$$u_{n_0} \in L(\infty)_{-\alpha+n_0\alpha_i},$$

and $f_i^{(n_0)} u_{n_0} = \tilde{f}_i^{n_0} u_{n_0} \in L(\infty)_{-\alpha}$ follows.

(step 3) $L_{-\alpha} = L(\infty)_{-\alpha} + vL_{-\alpha}$.

By (step 2), we have $L_{-\alpha} \subset L(\infty)_{-\alpha} + vL_{-\alpha}$. The opposite inclusion is obvious since $L(\infty) \subset L$ by (1).

Finally, we apply Nakayama's Lemma to the equality of (step 3) to obtain $L_{-\alpha} = L(\infty)_{-\alpha}$. □

Let B be the canonical basis and $L_A = \oplus_{b\in B}\mathbb{Z}[v]b$ as before. Note that we have already proved in Lemma 8.1 that $L(\infty) = L$ and that L is stable under $\tilde{e}_i$ and $\tilde{f}_i$. We first prove the following Lemma 8.2.

LEMMA 8.2. *Let $b \in B_{i,n}$. If we express $\pi_n^{-1}(b)$ as*

$$\pi_n^{-1}(b) = \sum_{k\geq 0} f_i^{(k)} u_k \quad (u_k \in \operatorname{Ker} e_i'),$$

then we have $u_k \in L_A$ and

$$b \equiv f_i^{(n)} u_0 \pmod{vL}, \quad \pi_n^{-1}(b) \equiv u_0 \pmod{vL}.$$

PROOF. Recall that $u_k, f_i^{(n+k)} u_k \in L_A$ by Lemma 7.8(6). Hence the element

$$b - f_i^{(n)} u_0 = b - f_i^{(n)} \pi_n^{-1}(b) + \sum_{k>0} \begin{bmatrix} n+k \\ k \end{bmatrix} f_i^{(n+k)} u_k$$

belongs to $\left(\sum_{k>0} f_i^{(n+k)} U_A^-\right) \cap L_A$ by Theorem 7.3(3) and Lemma 7.6(1). Since Lemma 7.6(1) implies that

$$\left(\sum_{k>0} f_i^{(n+k)} U_A^-\right) \cap L_A = \bigoplus_{b'\in B_{i,m}, m>n} \mathbb{Z}[v]b',$$

we have

$$b - f_i^{(n)} u_0 \in \bigoplus_{b'\in B_{i,m}, m>n} \mathbb{Z}[v]b'.$$

We now write

$$f_i^{(n)} u_0 = b + \sum_{\substack{b'\in B_{i,m} \\ m>n}} c_{b'}(v) b' \quad (c_{b'}(v) \in \mathbb{Z}[v])$$

and denote the constant term of $c_{b'}(v)$ by $c_{b'}$. Then we have

$$\begin{cases} (\, u_0,\, u_0\,) \equiv 1 + \displaystyle\sum_{b' \in B_{i,m}, m>n} c_{b'}^2 \pmod{v\mathbb{Z}[[v]]}, \\ \displaystyle\sum_{k\geq 0} (\, u_k,\, u_k\,) \equiv (\, \pi_n^{-1}(b),\, \pi_n^{-1}(b)\,) \equiv 1 \pmod{v\mathbb{Z}[[v]]}. \end{cases}$$

Applying Lemma 7.7(2) to $u_k \in L_A$, we obtain

$$(\, u_0,\, u_0\,) \equiv 1, \quad (\, u_k,\, u_k\,) \equiv 0 \quad (k \geq 1), \quad c_{b'}(v) \equiv 0.$$

Therefore, we have $f_i^{(n)} u_0 \equiv b \pmod{vL}$. Since $u_k \in vL$ $(k \geq 1)$ by Lemma 7.8(5'), we also have $\pi_n^{-1}(b) \equiv u_0 \pmod{vL}$. □

LEMMA 8.3 ([**Lusztig**, 19.2.1]). *Define* $B(\infty)$ *by*

$$B(\infty) = \left\{ \tilde{f}_{i_1} \cdots \tilde{f}_{i_N} 1 \pmod{vL(\infty)} \,\middle|\, N \geq 0, (i_1, \ldots, i_N) \in (\mathbb{Z}/r\mathbb{Z})^N \right\}.$$

Then we have the following.

(1) $B(\infty) = \{b \pmod{vL}\}_{b\in B}$.
(2) $\tilde{e}_i B(\infty) \subset B(\infty) \cup \{0\}$ *and* $\tilde{f}_i B(\infty) \subset B(\infty)$.
(3) *If* $b \in B(\infty)$ *satisfies* $\tilde{e}_i b \in B(\infty)$, *then* $\tilde{f}_i \tilde{e}_i b = b$.

PROOF. (1) We first prove that $\{b \pmod{vL}\}_{b\in B_{-\alpha}} \subset B(\infty)_{-\alpha}$ by induction on $\mathrm{ht}(\alpha)$.

This is true if $\mathrm{ht}(\alpha) = 0$. Assume that $1 \neq b \in B_{-\alpha}$. Then, by Theorem 7.3(3) and (5), we may assume that $b \in B_{i,n} \cap B_{-\alpha}$ $(n \geq 1)$. If we write $\pi_n^{-1}(b) = \sum f_i^{(k)} u_k$, then Lemma 8.2 implies that

$$b \equiv f_i^{(n)} u_0 \pmod{vL}, \quad \pi_n^{-1}(b) \equiv u_0 \pmod{vL}.$$

If we apply the induction hypothesis to $u_0 \equiv \pi_n^{-1}(b) \in B_{-\alpha+n\alpha_i}$, then we have $u_0 \pmod{vL} \in B(\infty)$ and hence $b \pmod{vL} = \tilde{f}_i^n u_0 \pmod{vL} \in B(\infty)$. We have the desired inclusion.

Next we prove that $B(\infty)_{-\alpha} \subset \{b \pmod{vL}\}_{b\in B_{-\alpha}}$ by induction on $\mathrm{ht}(\alpha)$.

Let $\beta = \tilde{f}_i \tilde{f}_{i_2} \cdots \tilde{f}_{i_N} 1$. By the induction hypothesis, we may choose $b \in B$ such that $\tilde{f}_{i_2} \cdots \tilde{f}_{i_N} 1 \equiv b$. Assume that $b \in B_{i,n} \cap B_{-\alpha+\alpha_i}$ and write $\pi_n^{-1}(b) = \sum f_i^{(k)} u_k$ $(u_k \in \operatorname{Ker} e_i')$. Then Lemma 8.2 implies that $u_k \in L_A$ and

$$b \equiv f_i^{(n)} u_0 \pmod{vL}, \quad \pi_n^{-1}(b) \equiv u_0 \pmod{vL}.$$

Since Lemma 8.1(1) implies $\tilde{f}_i(b - f_i^{(n)} u_0) \in vL$, we have $\beta \equiv \tilde{f}_i b \equiv f_i^{(n+1)} u_0$. Further, Theorem 7.3(3) and Lemma 7.6(1) together with

$$f_i^{(n+1)} u_0 - f_i^{(n+1)} \pi_n^{-1}(b) = -\sum_{k>0} f_i^{(n+1)} f_i^{(k)} u_k$$

imply that

$$f_i^{(n+1)} \pi_n^{-1}(b) - \pi_{n+1}\left(\pi_n^{-1}(b)\right) \in \bigoplus_{\substack{b' \in B_{i,m} \cap B_{-\alpha} \\ m \geq n+2}} Ab'$$

and that

$$f_i^{(n+1)}u_0 - f_i^{(n+1)}\pi_n^{-1}(b) \in \bigoplus_{\substack{b' \in B_{i,m} \cap B_{-\alpha} \\ m \geq n+2}} Ab'.$$

Since we have $f_i^{(n+1)}u_0 \in L_A$ and $\pi_{n+1}\left(\pi_n^{-1}(b)\right) \in L_A$, we may conclude that

$$f_i^{(n+1)}u_0 - \pi_{n+1}\left(\pi_n^{-1}(b)\right) \in \bigoplus_{\substack{b' \in B_{i,m} \cap B_{-\alpha} \\ m \geq n+2}} \mathbb{Z}[v]b'.$$

It follows that $f_i^{(n+1)}u_0 \equiv \pi_{n+1}\left(\pi_n^{-1}(b)\right) \pmod{vL}$, since

$$(\,\pi_{n+1}(\pi_n^{-1}(b)),\, \pi_{n+1}(\pi_n^{-1}(b))\,) \equiv 1, \quad (\,f_i^{(n+1)}u_0,\, f_i^{(n+1)}u_0\,) \equiv 1.$$

Therefore, we have $\beta \equiv \pi_{n+1}\left(\pi_n^{-1}(b)\right) \pmod{vL}$. We have the desired inclusion.
(2)(3) Since the proofs are similar, we only prove $\tilde{e}_i B(\infty) \subset B(\infty) \cup \{0\}$. By (1), it is enough to show the statement for $b \pmod{vL}$ $(b \in B)$. Let $b \in B_{i,n}$ and express $\pi_n^{-1}(b)$ as $\pi_n^{-1}(b) = \sum f_i^{(k)}u_k$. Then Lemma 8.2 implies that

$$f_i^{(n)}u_0 \equiv b \pmod{vL}, \quad u_0 \equiv \pi_n^{-1}(b) \pmod{vL}.$$

Hence the following equation holds by Lemma 8.1(1).

$$\tilde{e}_i b \equiv \begin{cases} f_i^{(n-1)}u_0 & (n > 0) \\ 0 & (n = 0) \end{cases} \pmod{vL}$$

This proves (3). We now assume that $n > 0$, and prove that $\tilde{e}_i b \pmod{vL} \in B(\infty)$. In fact, the same argument as in (1) proves that

$$f_i^{(n-1)}u_0 - \pi_{n-1}\left(\pi_n^{-1}(b)\right) \in \bigoplus_{\substack{b' \in B_{i,m} \cap B_{-\alpha} \\ m \geq n}} \mathbb{Z}[v]b',$$

and if we combine it with

$$(\,f_i^{(n-1)}u_0,\, f_i^{(n-1)}u_0\,) \equiv 1, \quad (\,\pi_{n-1}(\pi_n^{-1}(b)),\, \pi_{n-1}(\pi_n^{-1}(b))\,) \equiv 1,$$

then we have $f_i^{(n-1)}u_0 \equiv \pi_{n-1}\left(\pi_n^{-1}(b)\right) \pmod{vL}$. Since (1) says that

$$\pi_{n-1}(\pi_n^{-1}(b)) \pmod{vL} \in B(\infty),$$

we have $\tilde{e}_i b \pmod{vL} \in B(\infty)$. □

LEMMA 8.4. *Let (L, B) be a crystal basis of U_v^-. Then we have the following.*

(1) $L = \bigoplus_{n \geq 0} f_i^{(n)}(\,L \cap \operatorname{Ker} e_i'\,)$.

(2) $L^\vee = \{u \in U_v^- \mid (\,u,\, L\,) \subset \mathbb{Q}[[v]]\}$ *is stable under the Kashiwara operators $\tilde{e}_i$ and $\tilde{f}_i$.*

PROOF. (1) Take $u \in L$ and write $u = \sum_{n=0}^N f_i^{(n)}u_n$ $(u_n \in \operatorname{Ker} e_i')$. We prove $L \subset \oplus_{n \geq 0} f_i^{(n)}(\,L \cap \operatorname{Ker} e_i'\,)$ by induction on N.

As $\tilde{e}_i' L \subset L$, we may apply the induction hypothesis to $\tilde{e}_i' u$ and conclude that $u_n \in L \cap \operatorname{Ker} e_i'$ $(n \geq 1)$. Using $\tilde{f}_i L \subset L$, we also have $u_0 \in L \cap \operatorname{Ker} e_i'$. The opposite inclusion follows from $\tilde{f}_i L \subset L$.
(2) Set $L_n = f_i^{(n)}(\,L \cap \operatorname{Ker} e_i'\,)$ and $L_n^\vee = \{u \in f_i^{(n)} \operatorname{Ker} e_i' \mid (\,u,\, L_n\,) \subset \mathbb{Q}[[v]]\}$. Then we have $L^\vee = \oplus L_n^\vee$ by (1) and Lemma 7.8(3).

We shall show that $\tilde{e}_i L_n^{\vee} \subset L_{n-1}^{\vee}$. Let $f_i^{(n)}u$ be an element of $L_n^{\vee}$. Then for any element $f_i^{(n-1)}u' \in L_{n-1}$, where $u' \in L \cap \operatorname{Ker} e_i'$, we have

$$(\, \tilde{e}_i f_i^{(n)} u,\, f_i^{(n-1)} u' \,) = (1 - v^{2n})(\, f_i^{(n)} u,\, f_i^{(n)} u' \,)$$

by Lemma 7.7(1). Hence $f_i^{(n)}u' \in L_n$ implies that

$$(\, \tilde{e}_i f_i^{(n)} u,\, L_{n-1} \,) \subset \mathbb{Q}[[v]].$$

We have proved $\tilde{e}_i L_n^{\vee} \subset L_{n-1}^{\vee}$. The proof that $\tilde{f}_i L_n^{\vee} \subset L_{n+1}^{\vee}$ is similar. □

8.2. The first main theorem

We are now in a position to state the first main theorem. This theorem is a general result (see [**Kashiwara**, Theorem 4]), but we prove it only in the $A_{r-1}^{(1)}$ case. The proof we adopt is due to Lusztig [**Lusztig**, 19.2.1]. It was originally proved in [**cb-GL2**], generalizing Lusztig's proof for the quantum algebras of finite type. (For example, A_{r-1} is of finite type.)

THEOREM 8.5. **(The first main theorem of Kashiwara and Lusztig)** *We have the following existence and uniqueness of the crystal basis of U_v^-.*

(1) *$(L(\infty), B(\infty))$ is a crystal basis of U_v^-.*
(2) *Any crystal basis of U_v^- coincides with $(L(\infty), B(\infty))$ up to a constant multiple.*

PROOF. (1) $L(\infty) = L$ is an R-lattice of U_v^- by Lemma 8.1(2), and $B(\infty) = \{b \pmod{vL}\}_{b \in B}$ is a $\mathbb{Q}$-basis of L/vL by Lemma 8.3(1). They both have root space decompositions. The remaining properties required are already proved in Lemma 8.1(1) and Lemma 8.3(2)(3).
(2) Let (L, B) be a crystal basis. By multiplying by a suitable scalar, we may assume that $L \cap K = R$ and $1 \pmod{vL} \in B$. Now $1 \in L^{\vee}$ since $(\,1, 1\,) = 1$, so Lemma 8.4(2) implies that $L(\infty) \subset L^{\vee}$. Hence we have $L \subset L^{\vee\vee} \subset L(\infty)^{\vee}$. Combined with the inclusion $L(\infty) \subset L$, we conclude that $L(\infty) \subset L \subset L(\infty)^{\vee}$.

Hence it is enough to show $L(\infty)^{\vee} \subset L(\infty)$. To prove this, take $u \in L(\infty)^{\vee}$ and write u as a linear combination of elements of the canonical basis as follows.

$$u = \sum c_b(v) b \quad (c_b(v) \in K)$$

We consider K as a subfield of $\mathbb{Q}((v))$ and expand the coefficients $c_b(v)$:

$$c_b(v) = r_b v^{l_b} + (\textit{higher degree terms}) \quad (r_b \in \mathbb{Q}^{\times}).$$

(If $c_b(v) = 0$ we set $l_b = \infty$.) Let l be the minimum of the l_b and fix b which attains the minimum l. Then we have

$$(\, u,\, b \,) = r_b v^l + (\textit{higher degree terms}) \in \mathbb{Q}[[v]],$$

which implies $l \geq 0$. In particular, we have $l_b \geq 0$ and $u \in L(\infty)$. Hence we have $L(\infty)^{\vee} \subset L(\infty)$, and $L = L(\infty)$ follows.

Next we prove that B coincides with $B(\infty)$. The inclusion $B(\infty) \subset B$ is obvious since $1 \pmod{vL} \in B$. As $B(\infty)$ and B are both bases of L/vL, this implies that $B = B(\infty)$. □

We can characterize the canonical basis using the crystal basis $(L(\infty), B(\infty))$.

THEOREM 8.6. **(Characterization of the canonical basis)**
The canonical basis is the unique basis B of U_v^- such that

(1) $B \subset L(\infty)$,
(2) $\bar{b} = b \ (b \in B)$,
(3) $B(\infty) = \{b \ (\mathrm{mod}\ vL(\infty))\}_{b \in B}$.

PROOF. The canonical basis satisfies the properties (1), (2) and (3) by Theorem 7.3(2), Lemma 8.1 and Lemma 8.3. To show uniqueness, we assume that two bases B_1 and B_2 satisfy the three conditions above. Set $L_1 = \sum_{b_1 \in B_1} Rb_1$ and $L_2 = \sum_{b_2 \in B_2} Rb_2$. Then properties (1) and (3) imply

$$L_1 + vL(\infty) = L(\infty), \quad L_2 + vL(\infty) = L(\infty).$$

Hence B_1 and B_2 are R-free bases of $L(\infty)$ by Nakayama's Lemma.

For each $b_1 \in B_1$, we may choose $b_2 \in B_2$ satisfying $b_1 \equiv b_2 \ (\mathrm{mod}\ vL(\infty))$ by property (3). We now express their difference in terms of the canonical basis as follows.

$$b_1 - b_2 = \sum c_b(v)b \quad (c_b(v) \in vR).$$

Then property (2) implies that

$$b_1 - b_2 = \bar{b}_1 - \bar{b}_2 = \sum c_b(v^{-1})b.$$

Since $c_b(v) = c_b(v^{-1}) \in vR$ is only possible when $c_b(v) = 0$, we have $b_1 = b_2$. We have proved that $B_1 \subset B_2$. By interchanging the roles of B_1 and B_2, we have the opposite inclusion. □

CHAPTER 9

Existence and uniqueness (part II)

9.1. Preparatory results

In this chapter, we prove the existence and uniqueness of the crystal bases of integrable U_v-modules. Before proving the result, we prepare several results in this section. As in the previous chapters, $[k] = \frac{v^k - v^{-k}}{v - v^{-1}}$.

PROPOSITION 9.1. *Let $V(\lambda)$ be an integrable highest weight U_v-module, v_λ its highest weight vector. We denote by B the canonical basis of U_v^-. Then we have the following.*

(1) *We have an isomorphism of U_v^--modules*

$$V(\lambda) \simeq U_v^- / \sum_{i=0}^{r-1} U_v^- f_i^{1+\lambda(h_i)}.$$

(2) *$\{bv_\lambda \mid b \in B\} \setminus \{0\}$ is a basis of $V(\lambda)$.*

PROOF. (1) The following formulas may be proved by induction on k.

$$[e_i, f_i^k] = \frac{v^{-k+1}[k]}{v - v^{-1}} f_i^{k-1} t_i - \frac{v^{k-1}[k]}{v - v^{-1}} f_i^{k-1} t_i^{-1}.$$

We define I'_λ by

$$I'_\lambda = \left(\sum_{i=0}^{r-1} U_v^- f_i^{1+\lambda(h_i)} \right) U_v^0 U_v^+ + U_v^- U_v^0 \left(\sum_{i=0}^{r-1} U_v^+ e_i \right)$$
$$+ U_v^- \left(U_v^0 (v^d - v^{\lambda(d)}) + \sum_{i=0}^{r-1} U_v^0 (t_i - v^{\lambda(h_i)}) \right) U_v^+.$$

Then the following equations prove that I'_λ is a left ideal of U_v.

$$e_j f_i^{1+\lambda(h_i)} = f_i^{1+\lambda(h_i)} e_j + \delta_{ij} \frac{v^{-\lambda(h_i)}[\lambda(h_i)+1]}{v - v^{-1}} f_i^{\lambda(h_i)} (t_i - v^{\lambda(h_i)})$$
$$- \delta_{ij} \frac{v^{\lambda(h_i)}[\lambda(h_i)+1]}{v - v^{-1}} f_i^{\lambda(h_i)} (t_i^{-1} - v^{-\lambda(h_i)}),$$

$$e_j (t_i - v^{\lambda(h_i)}) = v^{-\alpha_j(h_i)} (t_i - v^{\lambda(h_i)}) e_j + v^{\lambda(h_i)} (v^{-\alpha_j(h_i)} - 1) e_j,$$
$$e_j (v^d - v^{\lambda(d)}) = v^{-\delta_{j0}} (v^d - v^{\lambda(d)}) e_j + v^{\lambda(d)} (v^{-\delta_{j0}} - 1) e_j.$$

Let I_λ be the left ideal of U_v generated by $\{e_i, f_i^{1+\lambda(h_i)}, t_i - v^{\lambda(h_i)}\}_{0 \le i < r}$ and $v^d - v^{\lambda(d)}$. Then we have $V(\lambda) = U_v / I_\lambda$, by definition, and $I'_\lambda \supset I_\lambda$ since I'_λ contains all of the generators of I_λ. Thus we have a surjective U_v-module homomorphism

$V(\lambda) \to U_v/I'_\lambda$. By the triangular decomposition of U_v, U_v/I'_λ is isomorphic to $U_v^- / \sum_{i=0}^{r-1} U_v^- f_i^{1+\lambda(h_i)}$ as a vector space. Recalling that $V(\lambda)$ is irreducible by Lemma 6.9(2), the homomorphism $V(\lambda) \to U_v/I'_\lambda$ must be an isomorphism, since U_v/I'_λ is non-zero.

(2) By Lemma 7.6(1), $f_i^{1+\lambda(h_i)} U_v^-$ has $\sqcup_{m \geq 1+\lambda(h_i)} B_{i,m}$ as a basis. Hence a subset of B gives a basis of $\sum_{i=0}^{r-1} f_i^{1+\lambda(h_i)} U_v^-$. Applying the antiautomorphism σ in Theorem 7.3(6), we know that a subset of B gives a basis of $\sum_{i=0}^{r-1} U_v^- f_i^{1+\lambda(h_i)}$. Let $B^c[\lambda]$ be its complement in B. Then for $b \in B$, we have

$$bv_\lambda = 0 \Leftrightarrow b \in \sum_{i=0}^{r-1} U_v^- f_i^{1+\lambda(h_i)} \Leftrightarrow b \notin B^c[\lambda].$$

Hence, we have that $\{bv_\lambda \mid b \in B\} \setminus \{0\}$ equals $\{bv_\lambda\}_{b \in B^c[\lambda]}$. We now prove that $\{bv_\lambda\}_{b \in B^c[\lambda]}$ is a basis of $V(\lambda)$. It obviously spans $V(\lambda)$. Assume that

$$\sum_{bv_\lambda \neq 0} c_b(v) bv_\lambda = 0.$$

Then we have $\sum_{b \in B^c[\lambda]} c_b(v) b \in \sum_{b \notin B^c[\lambda]} Kb$. This implies that $c_b(v) = 0$. Hence, $\{bv_\lambda \neq 0\}$ is linearly independent. □

LEMMA 9.2. ([**Kashiwara**, Remark 3.4.11] [**Lusztig**, 17.1.2])
Let $U_v \otimes_K \mathcal{B}_v$ be the tensor product of K-algebras and let $\Delta'(e'_i)$ and $\Delta'(f_i)$ be the elements defined by

$$\begin{aligned} \Delta'(f_i) &= f_i \otimes 1 + t_i \otimes f_i, \\ \Delta'(e'_i) &= -(v - v^{-1}) t_i e_i \otimes 1 + t_i \otimes e'_i. \end{aligned}$$

Then we have the following.

(1) *There exists a unique K-algebra homomorphism $\Delta' : \mathcal{B}_v \to U_v \otimes \mathcal{B}_v$ which extends $\Delta'(f_i), \Delta'(e'_i)$ as defined above.*
(2) $(\Delta \otimes \mathrm{id}) \circ \Delta' = (\mathrm{id} \otimes \Delta') \circ \Delta'$.
(3) *Consider $V(\lambda) \otimes U_v^-$ as a $\mathcal{B}_v$-module by using the algebra homomorphism Δ'. Then we have*

$$V(\lambda) \otimes U_v^- = \bigoplus_{n \geq 0} f_i^{(n)} \operatorname{Ker} e'_i.$$

In particular, the Kashiwara operators $\tilde{e}_i, \tilde{f}_i$ are well-defined on $V(\lambda) \otimes U_v^-$.

PROOF. (1) It is enough to check the defining relations of $\mathcal{B}_v$, which is straightforward.
(2) It is enough to check the equation on the generators f_i and e'_i.
(3) We use $P_i = \sum (-1)^n v^{-n(n-1)/2} f_i^{(n)} e_i'^n$ and follow the proof of Proposition 6.14. □

LEMMA 9.3 ([**Lusztig**, Proposition 17.1.13]). *Let V_l be the representation of $U_v(\mathfrak{sl}_2)$ and (L_l, B_l) its crystal basis as defined in Lemma 5.1. We view $U_v^-(\mathfrak{sl}_2)$ as a representation of $\mathcal{B}_v(\mathfrak{sl}_2)$. Then the following hold for the tensor product $V_l \otimes_K U_v^-(\mathfrak{sl}_2)$.*

(1) *Set $L := \sum_{n \geq 0} R f^{(n)}$. Then $L_l \otimes L$ is stable under the Kashiwara operators $\tilde{e}$ and $\tilde{f}$.*

(2) *Set* $B := \{f^{(n)} \pmod{vL}\}_{n\geq 0}$. *Then* $B_l \times B$ *has the property*

$$\tilde{e}\,(B_l \times B) \subset (B_l \times B) \cup \{0\}, \quad \tilde{f}\,(B_l \times B) \subset B_l \times B,$$

and $\tilde{e}b \neq 0$ *implies* $b = \tilde{f}\tilde{e}b$. *Further, the action of* $\tilde{f}$ *is given by the following rule.*

$$\tilde{f}\left(u_i \otimes f^{(j)}\right) \equiv \begin{cases} u_{i+1} \otimes f^{(j)} & (i+j<l) \\ u_i \otimes f^{(j+1)} & (i+j \geq l) \end{cases} \pmod{vL_l \otimes L}$$

PROOF. We introduce symmetric bilinear forms on V_l and U_v^- as follows.

$$(\, u_i,\, u_j\,) = \delta_{ij}\frac{(1-v^{2l})\cdots(1-v^{2l-2i+2})}{(1-v^{2i})\cdots(1-v^2)},$$

$$(\, f^{(i)},\, f^{(j)}\,) = \delta_{ij}\frac{1}{(1-v^{2i})\cdots(1-v^2)}.$$

Then we can check that the following formulas hold.

$$(9.1)\qquad \begin{cases} (\, fu_i,\, u_j\,) = v^{-1}(\, u_i,\, teu_j\,), \quad (\, tu_i,\, u_j\,) = (\, u_i,\, tu_j\,) \\ (\, ff^{(i)},\, f^{(j)}\,) = \dfrac{1}{1-v^2}(\, f^{(i)},\, e'f^{(j)}\,) \end{cases}$$

We define a symmetric bilinear form on $V_l \otimes U_v^-$ by

$$(\, u_{i_1} \otimes f^{(j_1)},\, u_{i_2} \otimes f^{(j_2)}\,) = (\, u_{i_1},\, u_{i_2}\,)(\, f^{(j_1)},\, f^{(j_2)}\,).$$

Then the elements $b_{ij} := u_i \otimes f^{(j)}$ satisfy

$$(\, b_{i_1j_1},\, b_{i_2j_2}\,) \in \delta_{i_1i_2}\delta_{j_1j_2} + v\mathbb{Z}[[v]].$$

Thus (9.1) implies that

$$(\, fb_{i_1j_1},\, b_{i_2j_2}\,) = \frac{1}{1-v^2}(\, b_{i_1j_1},\, e'b_{i_2j_2}\,).$$

We define elements v_k for $0 \leq k \leq l$ by

$$v_k = \sum_{i=0}^{k} \frac{(-1)^i v^{(l-k+1)i}}{(1-v^{2l})\cdots(1-v^{2l-2i+2})} u_i \otimes f^{(k-i)}.$$

Then $v_k \in \operatorname{Ker} e'$ and $(\, v_k,\, v_k\,) \in 1 + v\mathbb{Z}[[v]]$. We also have

$$(\, f^{(n)}v_k,\, f^{(n)}v_k\,) \in 1 + v\mathbb{Z}[[v]]$$

by induction on n.

Now Lemma 9.3 is proved by the following four assertions.

Assertion 1 *If we write* $f^{(n)}v_k = \sum_{0\leq i\leq l}\sum_{j\geq 0} c_{ij}(v)b_{ij}$ $(c_{ij}(v) \in K)$, *then we have* $c_{ij}(v) \in \mathbb{Z}[[v]]$, *and there exists an index* (i_0, j_0) *such that*

$$c_{i_0j_0}(v) \equiv \pm 1 \pmod{v}, \quad c_{ij}(v) \equiv 0 \pmod{v} \quad ((i,j) \neq (i_0, j_0)).$$

First, we expand all $c_{ij}(v)$ as Laurent series. We may prove by induction on n that the coefficients are integers, and the degrees are bounded from below. Let $-l$ be the minimal degree. This is independent of i, j and we may write

$$c_{ij}(v) = n_{ij}v^{-l} + (\textit{higher degree terms}) \quad (n_{ij} \in \mathbb{Z})$$

such that $\sum n_{ij}^2 > 0$. If we substitute

$$f^{(n)}v_k = \sum_{0\le i\le l}\sum_{j\ge 0} c_{ij}(v)b_{ij}$$

into $(\, f^{(n)}v_k,\, f^{(n)}v_k\,) \equiv 1$, then we have

$$\sum n_{ij}^2 = 1, \quad l = 0$$

since $(\, b_{i_1j_1},\, b_{i_2j_2}\,) \in \delta_{i_1i_2}\delta_{j_1j_2} + v\mathbb{Z}[[v]]$. In particular, we have $c_{ij}(v) \in \mathbb{Z}[[v]]$, and $c_{ij}(v) \equiv 0 \pmod v$ except for an index (i_0, j_0).

Assertion 2 $f^{(n)}v_k \equiv \begin{cases} u_n \otimes f^{(k)} & (0 \le n \le l-k) \\ u_{l-k} \otimes f^{(n-l+2k)} & (n \ge l-k) \end{cases} \pmod{vL_l \otimes L}.$

By Assertion 1 it is enough to find an index (i,j) with $c_{ij}(v) \equiv 1$. Since Lemma 5.3(1) and Lemma 6.12(3) imply that

$$\Delta'(f^{(n)}) = \sum_{j=0}^{n} v^{-j(n-j)} f^{(j)} t^{n-j} \otimes f^{(n-j)},$$

we have

$$(9.2) \quad f^{(n)}v_k = \sum_{i=0}^{k}\sum_{j=0}^{n} \frac{(-1)^i v^{(l-k+1)i+(n-j)(l-2i-j)}}{(1-v^{2l})\cdots(1-v^{2l-2i+2})} \times \begin{bmatrix} i+j \\ j \end{bmatrix} \begin{bmatrix} n+k-i-j \\ k-i \end{bmatrix} u_{i+j} \otimes f^{(n+k-i-j)}.$$

If $n \le l-k$ then the coefficient of $u_n \otimes f^{(k)}$ in (9.2) is

$$\sum_{i=0}^{\min(n,k)} (-1)^i v^{(2l-2k-2n+i+1)i} \equiv 1 \pmod v.$$

If $n \ge l-k$ then the coefficient of $u_{l-k} \otimes f^{(n-l+2k)}$ in (9.2) is

$$\sum_{i=0}^{\min(l-k,k)} (-1)^i v^{i(i+1)} \equiv 1 \pmod v.$$

Hence we have the result.

Assertion 3 *We have* $L_l \otimes L = \bigoplus_{k=0}^{l}\left(\underset{n\ge 0}{\oplus} Rf^{(n)}v_k\right)$. *In particular,* $L_l \otimes L$ *is stable under the Kashiwara operators* $\tilde{e}$ *and* $\tilde{f}$.

We have $f^{(n)}v_k \in L_l \otimes L$ by Assertion 1. Combined with Assertion 2 this implies that

$$L_l \otimes L = \bigoplus_{k=0}^{l}\left(\underset{n\ge 0}{\oplus} Rf^{(n)}v_k\right) + vL_l \otimes L.$$

We apply Nakayama's Lemma to each of the weight spaces and the result follows.

Assertion 4 $B_l \times B = \{b_{ij} \pmod{vL_l \otimes L}\}$ *has the following properties.*

(1) $\tilde{e}(B_l \times B) \subset B_l \times B \cup \{0\}$ and $\tilde{f}(B_l \times B) \subset B_l \times B$.
(2) We have either $\tilde{e}b_{ij} \equiv 0 \pmod{vL_l \otimes L}$ or $b_{ij} \equiv \tilde{f}\tilde{e}b_{ij} \pmod{vL_l \otimes L}$.
(3) $\tilde{f}b_{ij} \equiv \begin{cases} b_{i+1,j} & (i+j<l) \\ b_{i,j+1} & (i+j \geq l) \end{cases} \pmod{vL_l \otimes L}$.

These are consequences of Assertion 2. Note that $vL_l \otimes L$ is stable under the operators $\tilde{e}$ and $\tilde{f}$ by Assertion 3. □

By Lemma 6.12(3) we can and do identify $\mathcal{B}_v^-$ with U_v^-. We define a K-linear map $\phi_\lambda : U_v^- \to V(\lambda) \otimes U_v^-$ by $\phi_\lambda(u) = \Delta'(u)(v_\lambda \otimes 1)$. Then we have the following lemma.

LEMMA 9.4 ([**Lusztig**, Lemma 18.1.4]). *Let $\phi_\lambda : U_v^- \to V(\lambda) \otimes U_v^-$ be as above. Then*

(1) $\phi_\lambda(u) - (uv_\lambda) \otimes 1 \in \oplus_{\alpha \neq 0} V(\lambda) \otimes U_{-\alpha}$.
(2) *The map ϕ_λ is a $\mathcal{B}_v$-module homomorphism.*
(3) $\phi_\lambda(\tilde{e}_i u) = \tilde{e}_i \phi_\lambda(u)$ *and* $\phi_\lambda(\tilde{f}_i u) = \tilde{f}_i \phi_\lambda(u)$.

PROOF. (1) It is enough to prove the statement for $u = f_{i_1} \cdots f_{i_N}$. Now $\Delta'(u)$ has the form $\Delta'(u) = u \otimes 1 + \cdots$ because $\Delta'(f_i) = f_i \otimes 1 + t_i \otimes f_i$, so we have $\phi_\lambda(u) \in (uv_\lambda) \otimes 1 + \oplus_{\alpha \neq 0} V(\lambda) \otimes U_{-\alpha}$.
(2) First, $\phi_\lambda(f_i u) = f_i \phi_\lambda(u)$ is obvious since Lemma 9.2(1) implies that

$$\phi_\lambda(f_i u) = \Delta'(f_i)\Delta'(u)(v_\lambda \otimes 1) = \Delta'(f_i)\phi_\lambda(u) = f_i \phi_\lambda(u).$$

Next, we prove that $\phi_\lambda(e_i' u) = e_i' \phi_\lambda(u)$. Assume that $\phi_\lambda(e_i' u) = e_i' \phi_\lambda(u)$ holds for all $u \in U_{-\alpha'}$ whenever $\mathrm{ht}(\alpha') < \mathrm{ht}(\alpha)$. Then, for $f_j u \in U_{-\alpha}$ we have that

$$\begin{aligned} \phi_\lambda(e_i'(f_j u)) &= v^{-\alpha_j(h_i)} \phi_\lambda(f_j e_i' u) + \delta_{ij} \phi_\lambda(u) \\ &= v^{-\alpha_j(h_i)} f_j \phi_\lambda(e_i' u) + \delta_{ij} \phi_\lambda(u). \end{aligned}$$

We apply the induction hypothesis to $e_i' u \in U_{-\alpha+\alpha_j}$ to obtain

$$\phi_\lambda(e_i'(f_j u)) = (v^{-\alpha_j(h_i)} f_j e_i' + \delta_{ij})\phi_\lambda(u) = e_i' f_j \phi_\lambda(u) = e_i' \phi_\lambda(f_j u).$$

As $U_{-\alpha} = \sum_{j=0}^{r-1} f_j U_{-\alpha+\alpha_j}$, the induction may be continued.
(3) Write $u = \sum_{n \geq 0} f_i^{(n)} u_n$ ($u_n \in \operatorname{Ker} e_i'$). Then we have

$$\phi_\lambda(u) = \sum_{n \geq 0} f_i^{(n)} \phi_\lambda(u_n) \quad (\phi_\lambda(u_n) \in \operatorname{Ker} e_i')$$

by (2). Hence the statement is obvious. □

9.2. The second main theorem

In this section, we prove the existence and uniqueness of the crystal basis of integrable highest weight U_v-modules. The argument here is taken from [**Lusztig**, Chap.18].

THEOREM 9.5 ([**Lusztig**, Theorem 18.3.8]). *Let B be the canonical basis. We define $L(\lambda)$ and $B(\lambda)$ for an integrable U_v-module $V(\lambda)$ as follows.*

$$\begin{aligned} L(\lambda) &= \sum R\tilde{f}_{i_1} \cdots \tilde{f}_{i_N} v_\lambda, \\ B(\lambda) &= \{\tilde{f}_{i_1} \cdots \tilde{f}_{i_N} v_\lambda \pmod{vL(\lambda)}\} \setminus \{0\}. \end{aligned}$$

Then we have the following.

(1) $L(\lambda) = L(\infty)v_\lambda$ *and* $B(\lambda) = \{bv_\lambda \pmod{vL(\lambda)} \mid b \in B, bv_\lambda \neq 0\}$.
(2) *If* $b \in B$ *satisfies* $bv_\lambda \neq 0$, *then*

$$\tilde{e}_i(bv_\lambda) \equiv (\tilde{e}_i b)v_\lambda \pmod{vL(\lambda)},$$
$$\tilde{f}_i(bv_\lambda) \equiv (\tilde{f}_i b)v_\lambda \pmod{vL(\lambda)}.$$

PROOF. These are proved by induction on $\mathrm{ht}(\lambda - \mu)$. We consider three statements $(1)_{\mu'}$, $(2)_{\mu'}$ and $(3)_{\mu'}$ given below, and assume that they hold for those μ' which satisfy $\mathrm{ht}(\lambda - \mu') < \mathrm{ht}(\lambda - \mu)$. Note that it is easy to see that they hold for $\mu' = \lambda$.

$(1)_{\mu'}$: $L(\lambda)_{\mu'} = L(\infty)_{\mu'-\lambda}v_\lambda$.
$(2)_{\mu'}$: If $b \in B_{\mu'-\lambda}$ satisfies $bv_\lambda \neq 0$, then

$$\tilde{e}_i(bv_\lambda) \equiv (\tilde{e}_i b)v_\lambda \pmod{vL(\lambda)},$$
$$\tilde{f}_i(bv_\lambda) \equiv (\tilde{f}_i b)v_\lambda \pmod{vL(\lambda)}.$$

$(3)_{\mu'}$: $B(\lambda)_{\mu'} = \{bv_\lambda \pmod{vL(\lambda)} \mid bv_\lambda \neq 0,\ b \in B_{\mu'-\lambda}\}$.

The proof of this proposition is carried out in eleven steps. Recall that we have proved in Lemma 8.1(2) and Lemma 8.3(1) that

$$\begin{cases} L(\infty) &= \sum R\tilde{f}_1 \cdots \tilde{f}_N 1 = \bigoplus_{b \in B} Rb, \\ B(\infty) &= \{\tilde{f}_1 \cdots \tilde{f}_N 1 \pmod{vL(\infty)}\} = \{b \pmod{vL(\infty)}\}_{b \in B}. \end{cases}$$

We freely use these results without reference.

(step 1) *We have* $L(\lambda)_{\mu'} = \sum_{n \geq 0} f_i^{(n)}(\, L(\lambda)_{\mu'+n\alpha_i} \cap \mathrm{Ker}\, e_i\,)$.

We show that $L(\lambda)_{\mu'} \subset \sum_{n\geq 0} f_i^{(n)}(\, L(\lambda)_{\mu'+n\alpha_i} \cap \mathrm{Ker}\, e_i\,)$. The opposite inclusion is obvious by the definition of $L(\lambda)$. To do this, we write $u \in L(\lambda)_{\mu'}$ as $u = \sum_{n=0}^N f_i^{(n)} u_n$ ($u_n \in \mathrm{Ker}\ e_i$) and prove that $u_n \in L(\lambda)$ by induction on N.

Since we have $u \in L(\infty)_{\mu'-\lambda}v_\lambda$ by $(1)_{\mu'}$, $(2)_{\mu'}$ implies that

$$\tilde{e}_i u \in L(\infty)_{\mu'-\lambda+\alpha_i}v_\lambda + vL(\lambda)_{\mu'+\alpha_i}.$$

Hence we get $\tilde{e}_i u \in L(\lambda)_{\mu'+\alpha_i}$ by $(1)_{\mu'+\alpha_i}$. Now the induction hypothesis is applicable and we get $u_n \in L(\lambda)$ $(n \geq 1)$. We also have $u_0 = u - \sum_{n\geq 1} f_i^{(n)} u_n \in L(\lambda)$.

(step 2) *If* $\mathrm{ht}(\lambda - \mu') < \mathrm{ht}(\lambda - \mu)$ *then we have the following inclusions.*

$$\begin{cases} \tilde{e}_i(L(\lambda) \otimes L(\infty))_{\mu'} \subset (L(\lambda) \otimes L(\infty))_{\mu'+\alpha_i} \\ \tilde{f}_i(L(\lambda) \otimes L(\infty))_{\mu'} \subset (L(\lambda) \otimes L(\infty))_{\mu'-\alpha_i} \\ \tilde{e}_i(L(\lambda)_{\mu'} \otimes L(\infty)_{\mu-\mu'}) \subset (L(\lambda) \otimes L(\infty))_{\mu+\alpha_i} \\ \tilde{f}_i(L(\lambda)_{\mu'} \otimes L(\infty)_{\mu-\mu'}) \subset (L(\lambda) \otimes L(\infty))_{\mu-\alpha_i} \\ \phi_\lambda(L(\infty)_{\mu'-\lambda}) \subset (L(\lambda) \otimes L(\infty))_{\mu'} \\ \phi_\lambda(L(\infty)_{\mu-\lambda}) \subset (L(\lambda) \otimes L(\infty))_{\mu} \end{cases}$$

Let $L^{hw}_{\geq\mu'} = \underset{\nu\in\mu'+Q_+}{\oplus} L(\lambda)_\nu \cap \operatorname{Ker} e_i$. We have

$$(L(\lambda)\otimes L(\infty))_{\mu'} = \left(\left(\sum_{n\geq 0} f_i^{(n)} L^{hw}_{\geq\mu'}\right) \bigotimes L(\infty)\right)_{\mu'} \tag{9.3}$$

by (step 1). Since Lemma 8.4(1) and Theorem 8.5 imply that

$$L(\infty) = \oplus_{n\geq 0} f_i^{(n)}(L(\infty)\cap \operatorname{Ker} e_i'),$$

we conclude using Lemma 9.3(1) that $\left(\sum_{n\geq 0} f_i^{(n)} L^{hw}_{\geq\mu'}\right)\otimes L(\infty)$ is stable under the Kashiwara operators $\tilde{e}_i$ and $\tilde{f}_i$. Hence the first two inclusions follow since

$$\sum_{n\geq 0} f_i^{(n)} L^{hw}_{\geq\mu'} \subset L(\lambda).$$

The proof of the second two inclusions is similar.

To prove the last two inclusions, we repeatedly apply Lemma 9.4(3) and the second inclusion to $\phi_\lambda(1) = v_\lambda\otimes 1 \in L(\lambda)\otimes L(\infty)$. We conclude that $\tilde{f}_{i_1}\cdots\tilde{f}_{i_N}1 \in L(\infty)_{\mu-\lambda}$ satisfies

$$\phi_\lambda(\tilde{f}_{i_1}\cdots\tilde{f}_{i_N}1) = \tilde{f}_{i_1}\cdots\tilde{f}_{i_N}\phi_\lambda(1) \in L(\lambda)\otimes L(\infty).$$

We have proved the last inclusion $\phi_\lambda(L(\infty)_{\mu-\lambda}) \subset (L(\lambda)\otimes L(\infty))_\mu$. If we replace μ by μ' and argue as before then we have $\phi_\lambda(L(\infty)_{\mu'-\lambda}) \subset (L(\lambda)\otimes L(\infty))_{\mu'}$.

(step 3) *Write $y \in (L(\lambda)\otimes L(\infty))_{\mu'}$ as $y = y_0\otimes 1 + y'$ in accordance with the direct sum decomposition*

$$(L(\lambda)_{\mu'}\otimes 1) \oplus \left(\underset{\alpha\neq 0}{\oplus} L(\lambda)_{\mu'+\alpha}\otimes L(\infty)_{-\alpha}\right),$$

and similarly write $z = \tilde{f}_i y$ as $z = z_0\otimes 1 + z'$. Then $z_0 \equiv \tilde{f}_i y_0 \pmod{vL(\lambda)_{\mu'-\alpha_i}}$.

In (step 2) we have proved (9.3). Hence it is enough to prove the assertion for $y \in (f_i^{(n)} L^{hw}_{\geq\mu'})_{\mu'+\alpha}\otimes L(\infty)_{-\alpha}$.

Note that $\bigoplus_{\alpha\neq 0} L(\lambda)_{\mu'+\alpha}\otimes L(\infty)_{-\alpha}$ is the direct sum of $\bigoplus_{n\geq 1} L(\lambda)_{\mu'+n\alpha_i}\otimes f_i^{(n)}$ and

$$\bigoplus_{\alpha\neq 0}\bigoplus_{n\geq 0}\left(L(\lambda)\otimes f_i^{(n)}(\,L(\infty)_{-\alpha}\cap \operatorname{Ker} e_i'\,)\right)_{\mu'}.$$

Hence Lemma 9.3(2) implies that the contribution of $\tilde{f}_i y'$ and $\tilde{f}_i(y_0\otimes 1)$ to the component $L(\lambda)_{\mu'}\otimes 1$ belong to $vL(\lambda)_{\mu'}\otimes 1$ and $(\tilde{f}_i y_0 + vL(\lambda)_{\mu'})\otimes 1$ respectively. Hence we have the result.

(step 4) *We have $(1)_\mu : L(\infty)_{\mu-\lambda}v_\lambda = L(\lambda)_\mu$.*

First we prove that $L(\lambda)_\mu \subset L(\infty)_{\mu-\lambda}v_\lambda + vL(\lambda)_\mu$. We have

$$L(\lambda)_\mu = \sum_{i=0}^{r-1}\tilde{f}_i L(\lambda)_{\mu+\alpha_i} = \sum_{i=0}^{r-1}\tilde{f}_i (L(\infty)v_\lambda)_{\mu+\alpha_i}$$

by $(1)_{\mu+\alpha_i}$. So applying $(2)_{\mu+\alpha_i}$ to the left hand side, we have the result.

Second we prove that $(L(\infty)v_\lambda)_\mu \subset L(\lambda)_\mu$. As $L(\infty)_{\mu-\lambda} = \sum\tilde{f}_i L(\infty)_{\mu-\lambda+\alpha_i}$, it is enough to prove that $(\tilde{f}_i b)v_\lambda \in L(\lambda)_\mu$ for $b \in B_{\mu-\lambda+\alpha_i}$. To do this, we apply (step 3) to $y = \phi_\lambda(b) \in (L(\lambda) \otimes L(\infty))_{\mu+\alpha_i}$. Since $z = \phi_\lambda(\tilde{f}_i b)$, we have $\tilde{f}_i(bv_\lambda) \equiv (\tilde{f}_i b)v_\lambda \pmod{vL(\lambda)}$ by Lemma 9.4(1). On the other hand, we already know

$$bv_\lambda \in L(\infty)_{\mu-\lambda+\alpha_i} v_\lambda = L(\lambda)_{\mu+\alpha_i}$$

by $(1)_{\mu+\alpha_i}$, so we have $\tilde{f}_i(bv_\lambda) \in L(\lambda)_\mu$. This implies that $(\tilde{f}_i b)v_\lambda \in L(\lambda)_\mu$. Thus we have obtained $L(\lambda)_\mu = L(\infty)_{\mu-\lambda} v_\lambda + vL(\lambda)_\mu$, and Nakayama's Lemma implies $(1)_\mu$.

(step 5) *If $b \in B_{\mu-\lambda}$ and $b \equiv \tilde{f}_{i_1} \cdots \tilde{f}_{i_N} 1 \pmod{vL(\infty)}$ then*

$$bv_\lambda \equiv \tilde{f}_{i_1} \cdots \tilde{f}_{i_N} v_\lambda \pmod{vL(\lambda)}.$$

In particular, we have $(3)_\mu : \{bv_\lambda \pmod{vL(\lambda)} \mid b \in B_{\mu-\lambda}, bv_\lambda \neq 0\} = B(\lambda)_\mu$, and $B(\lambda)_\mu$ is a $\mathbb{Q}$-basis of $L(\lambda)_\mu / vL(\lambda)_\mu$.

We consider the direct sum decomposition used in (step 3) and prove by downward induction on r that the $L(\lambda) \otimes 1$ component of $\phi_\lambda(\tilde{f}_{i_r} \cdots \tilde{f}_{i_N} 1)$ is equal to $(\tilde{f}_{i_r} \cdots \tilde{f}_{i_N} v_\lambda) \otimes 1$ modulo $vL(\lambda) \otimes L(\infty)$. Assume that this is true for $r+1$. We express $\phi_\lambda(\tilde{f}_{i_{r+1}} \cdots \tilde{f}_{i_N} 1)$ as $y_{r+1} + y'_{r+1}$ in accordance with the direct sum decomposition of (step 3). Then $y_{r+1} - (\tilde{f}_{i_{r+1}} \cdots \tilde{f}_{i_N} v_\lambda) \otimes 1$ belongs to $vL(\lambda) \otimes L(\infty)$ by the induction hypothesis. Thus the second inclusion of (step 2) implies that

$$\tilde{f}_{i_r} y_{r+1} \equiv \tilde{f}_{i_r}\left((\tilde{f}_{i_{r+1}} \cdots \tilde{f}_{i_N} v_\lambda) \otimes 1\right) \pmod{vL(\lambda) \otimes L(\infty)}.$$

In particular, the $L(\lambda) \otimes 1$ component of $\tilde{f}_{i_r} y_{r+1}$ and the $L(\lambda) \otimes 1$ component of $\tilde{f}_{i_r}((\tilde{f}_{i_{r+1}} \cdots \tilde{f}_{i_N} v_\lambda) \otimes 1)$ are the same modulo $vL(\lambda) \otimes 1$. Applying (step 3), we know that the $L(\lambda) \otimes 1$ component of $\tilde{f}_{i_r}((\tilde{f}_{i_{r+1}} \cdots \tilde{f}_{i_N} v_\lambda) \otimes 1)$ is $\tilde{f}_{i_r} \cdots \tilde{f}_{i_N} v_\lambda \otimes 1$ modulo $vL(\lambda) \otimes 1$. On the other hand, the $L(\lambda) \otimes 1$ component of $\tilde{f}_{i_r} y_{r+1}$ is y_r, and the induction may be continued.

Now we apply the sixth inclusion of (step 2) to the element $b - \tilde{f}_{i_1} \cdots \tilde{f}_{i_N} 1 \in vL(\lambda) \otimes L(\infty)$. Then we have

$$\phi_\lambda(b) \equiv \phi_\lambda(\tilde{f}_{i_1} \cdots \tilde{f}_{i_N} 1) \pmod{vL(\lambda) \otimes L(\infty)}.$$

Taking the $L(\lambda) \otimes 1$ components of these, we get

$$bv_\lambda \equiv \tilde{f}_{i_1} \cdots \tilde{f}_{i_N} v_\lambda \pmod{vL(\lambda)}.$$

Before proving the rest of the statement, we prove here that if $b \in B_{\mu-\lambda}$ and $bv_\lambda \neq 0$, then $bv_\lambda \not\equiv 0 \pmod{vL(\lambda)}$. By (step 4) we have

$$L(\lambda)_\mu = \sum_{b \in B_{\mu-\lambda}, bv_\lambda \neq 0} R\, bv_\lambda,$$

and this sum is direct by Proposition 9.1. Thus we have

$$\bigoplus_{bv_\lambda \neq 0} \mathbb{Q}\, bv_\lambda \simeq L(\lambda)_\mu / vL(\lambda)_\mu$$

by the map $bv_\lambda \mapsto bv_\lambda \pmod{vL(\lambda)}$. Hence we get $bv_\lambda \not\equiv 0 \pmod{vL(\lambda)}$.

Now we may prove $(3)_\mu$. Let $b \in B_{\mu-\lambda}$ satisfy $bv_\lambda \neq 0$, and we write

$$b \equiv \tilde{f}_{i_1} \cdots \tilde{f}_{i_N} 1 \pmod{vL(\infty)}.$$

Then we have $bv_\lambda \not\equiv 0 \pmod{vL(\lambda)}$ and $bv_\lambda \equiv \tilde{f}_{i_1} \cdots \tilde{f}_{i_N} v_\lambda \pmod{vL(\lambda)}$. Hence $bv_\lambda \pmod{vL(\lambda)} \in B(\lambda)_\mu$ and we have proved the inclusion

$$\{bv_\lambda \pmod{vL(\lambda)} \mid b \in B_{\mu-\lambda}, bv_\lambda \neq 0\} \subset B(\lambda)_\mu.$$

To prove the opposite inclusion, we take an element $\tilde{f}_{i_1} \cdots \tilde{f}_{i_N} v_\lambda \pmod{vL(\lambda)}$ of $B(\lambda)_\mu$. Then there exists $b \in B_{\mu-\lambda}$ such that $b \equiv \tilde{f}_{i_1} \cdots \tilde{f}_{i_N} 1 \pmod{vL(\infty)}$. Since this b satisfies $bv_\lambda \equiv \tilde{f}_{i_1} \cdots \tilde{f}_{i_N} v_\lambda \not\equiv 0$, we have $bv_\lambda \neq 0$.

(step 6) *For $b \in B_{\mu'-\lambda}$, we write $bv_\lambda = \sum_{n=0}^{N} f_i^{(n)} u_n$. Then we have $u_n \in L(\lambda)$ for all n, and there exists an integer n_0 such that $bv_\lambda \equiv f_i^{(n_0)} u_{n_0} \pmod{vL(\lambda)}$.*

By $(1)_{\mu'}$ and (step 1) we have $u_n \in L(\lambda)$. Let n_0 be the smallest n which satisfies $\tilde{e}_i^{n+1}(bv_\lambda) \equiv 0 \pmod{vL(\lambda)}$. Then we claim that there exists a sequence of elements $b_k \in B$ $(0 \leq k \leq n_0)$ such that

$$\tilde{e}_i^k(bv_\lambda) \equiv b_k v_\lambda \pmod{vL(\lambda)}, \quad \tilde{e}_i b_k \equiv b_{k+1} \pmod{vL(\infty)}.$$

To prove this, we set $b_0 = b$ for $k = 0$ and assume that b_k exists. Then we get $\tilde{e}_i(b_k v_\lambda) \equiv (\tilde{e}_i b_k) v_\lambda$ by $b_k v_\lambda \neq 0$ and $(2)_{\mu'+k\alpha_i}$. Since $\tilde{e}_i(b_k v_\lambda) \not\equiv 0 \pmod{vL(\lambda)}$ and $(1)_{\mu'+(k+1)\alpha_i}$ imply that $\tilde{e}_i b_k \not\equiv 0 \pmod{vL(\infty)}$, Lemma 8.3(2) guarantees the existence of $b_{k+1} \in B$ with $b_{k+1} \equiv \tilde{e}_i b_k \pmod{vL(\infty)}$. Hence the induction may be continued.

Recalling properties of $B(\infty)$, $(2)_{\mu'+k\alpha_i}$ and $b_{k-1} \equiv \tilde{f}_i \tilde{e}_i b_{k-1} \pmod{vL(\infty)}$ for $1 \leq k \leq n_0$ imply that

$$\tilde{f}_i \tilde{e}_i^k(bv_\lambda) \equiv \tilde{f}_i(b_k v_\lambda) \equiv (\tilde{f}_i b_k) v_\lambda \equiv b_{k-1} v_\lambda \equiv \tilde{e}_i^{k-1}(bv_\lambda) \pmod{vL(\lambda)}.$$

This means that

$$\sum_{n=k}^{N} f_i^{(n-k+1)} u_n \equiv \sum_{n=k-1}^{N} f_i^{(n-k+1)} u_n \pmod{vL(\lambda)},$$

so $u_0, f_i u_1, \ldots, f_i^{(n_0-1)} u_{n_0-1} \in vL(\lambda)$. If we apply (step 1) to

$$\tilde{e}_i^{n_0+1}(bv_\lambda) = \sum_{n=n_0+1}^{N} f_i^{(n-n_0-1)} u_n \in vL(\lambda),$$

we also have $u_{n_0+1}, \ldots, u_N \in vL(\lambda)$.

(step 7) *If $\tilde{f}_{i_1} \cdots \tilde{f}_{i_N} v_\lambda \in L(\lambda)_\mu$ and $\tilde{f}_{i_1} \cdots \tilde{f}_{i_N} v_\lambda \not\equiv 0 \pmod{vL(\lambda)}$ then*

$$\phi_\lambda(\tilde{f}_{i_1} \cdots \tilde{f}_{i_N} 1) \equiv (\tilde{f}_{i_1} \cdots \tilde{f}_{i_N} v_\lambda) \otimes 1 \pmod{vL(\lambda) \otimes L(\infty)}.$$

We prove the following by induction on r.

$$\phi_\lambda(\tilde{f}_{i_r} \cdots \tilde{f}_{i_N} 1) \equiv (\tilde{f}_{i_r} \cdots \tilde{f}_{i_N} v_\lambda) \otimes 1 \pmod{vL(\lambda) \otimes L(\infty)}.$$

Assume that this holds for $r+1$. If we have $\tilde{f}_{i_{r+1}} \cdots \tilde{f}_{i_N} 1 \in vL(\infty)$, then the fifth inclusion of (step 2) implies that

$$\phi_\lambda(\tilde{f}_{i_{r+1}} \cdots \tilde{f}_{i_N} 1) \equiv 0 \pmod{vL(\lambda) \otimes L(\infty)}.$$

Applying the induction hypothesis to the $L(\lambda) \otimes 1$ component, we obtain

$$\tilde{f}_{i_{r+1}} \cdots \tilde{f}_{i_N} v_\lambda \equiv 0 \pmod{vL(\lambda)},$$

which contradicts our assumption. Therefore, there exists $b \in B$ such that

$$\tilde{f}_{i_{r+1}} \cdots \tilde{f}_{i_N} 1 \equiv b \pmod{vL(\infty)}.$$

By (step 5), we have $bv_\lambda \equiv \tilde{f}_{i_{r+1}} \cdots \tilde{f}_{i_N} v_\lambda \pmod{vL(\lambda)}$, and the induction hypothesis implies that

$$\phi_\lambda(\tilde{f}_{i_{r+1}} \cdots \tilde{f}_{i_N} 1) \equiv bv_\lambda \otimes 1 \pmod{vL(\lambda) \otimes L(\infty)}.$$

We use (step 6) here to conclude that there exist an integer n_0 and $u_{n_0} \in L(\lambda)$ such that $bv_\lambda \equiv f_{i_r}^{(n_0)} u_{n_0} \pmod{vL(\lambda)}$. Hence we are in the situation that Lemma 9.3 is applicable. Since we have $f_{i_r}^{(n_0+1)} u_{n_0} \not\equiv 0$ by

$$\tilde{f}_{i_r} bv_\lambda \equiv \tilde{f}_{i_r} \tilde{f}_{i_{r+1}} \cdots \tilde{f}_{i_N} v_\lambda \not\equiv 0 \pmod{vL(\lambda)},$$

Lemma 9.3(2) implies that

$$\tilde{f}_{i_r}(bv_\lambda \otimes 1) \equiv (\tilde{f}_{i_r} bv_\lambda) \otimes 1 \pmod{vL(\lambda) \otimes L(\infty)}. \tag{9.4}$$

We apply $\tilde{f}_{i_r}$ to $\phi_\lambda(\tilde{f}_{i_{r+1}} \cdots \tilde{f}_{i_N} 1) \equiv bv_\lambda \otimes 1$ keeping (step 2) in mind. Then (9.4) implies that

$$\phi_\lambda(\tilde{f}_{i_r} \cdots \tilde{f}_{i_N} 1) \equiv (\tilde{f}_{i_r} \cdots \tilde{f}_{i_N} v_\lambda) \otimes 1 \pmod{vL(\lambda) \otimes L(\infty)}.$$

Hence the induction may be continued.

(step 8) *We have that* $\tilde{e}_i L(\lambda)_\mu \subset L(\lambda)_{\mu+\alpha_i}$. *Further, if* $\tilde{f}_{i_1} \cdots \tilde{f}_{i_N} v_\lambda \in L(\lambda)_\mu$ *satisfies* $\tilde{f}_{i_1} \cdots \tilde{f}_{i_N} v_\lambda \not\equiv 0 \pmod{vL(\lambda)}$ *then*

$$\phi_\lambda(\tilde{e}_i \tilde{f}_{i_1} \cdots \tilde{f}_{i_N} 1) \equiv (\tilde{e}_i \tilde{f}_{i_1} \cdots \tilde{f}_{i_N} v_\lambda) \otimes 1 \pmod{vL(\lambda) \otimes L(\infty)}. \tag{9.5}$$

We prove that $\tilde{e}_i L(\lambda)_\mu \subset v^{-n} L(\lambda)_{\mu+\alpha_i}$ implies $\tilde{e}_i L(\lambda)_\mu \subset v^{-n+1} L(\lambda)_{\mu+\alpha_i}$. To do this, we consider the collection of $u = \tilde{f}_{i_1} \cdots \tilde{f}_{i_N} v_\lambda \in L(\lambda)_\mu$ with $u \not\equiv 0 \pmod{vL(\lambda)}$. Then Nakayama's Lemma implies that these generate $L(\lambda)_\mu$ as an R-module. Thus it is enough to prove that $\tilde{e}_i u \in v^{-n+1} L(\lambda)_{\mu+\alpha_i}$ for u of this form. Now (step 7) implies that

$$\phi_\lambda(\tilde{f}_{i_1} \cdots \tilde{f}_{i_N} 1) \equiv u \otimes 1 \pmod{vL(\lambda) \otimes L(\infty)}.$$

So if we write $u = \sum_{k=0}^N f_i^{(k)} u_k$ and apply (step 1) to $\tilde{e}_i u \in v^{-n} L(\lambda)_{\mu+\alpha_i}$, then we get $u_k \in v^{-n} L(\lambda)$ $(k \geq 1)$ and

$$\tilde{e}_i(u \otimes 1) = \sum_{k=1}^N \tilde{e}_i(f_i^{(k)} u_k \otimes 1).$$

If we apply Lemma 9.3(2) to the tensor product

$$\left(\sum_{m \geq 0} R f_i^{(m)} u_k \right) \bigotimes \left(\sum_{m \geq 0} R f_i^{(m)} \right),$$

then we obtain

$$\tilde{e}_i(f_i^{(k)} u_k \otimes 1) - f_i^{(k-1)} u_k \otimes 1 \in v^{-n+1} L(\lambda) \otimes L(\infty).$$

Summing over k, we get

$$\tilde{e}_i(u \otimes 1) - (\tilde{e}_i u) \otimes 1 \in v^{-n+1} L(\lambda) \otimes L(\infty). \tag{9.6}$$

Since $\tilde{e}_i L(\lambda)_\mu \in v^{-n} L(\lambda)$, the above equation combined with the third inclusion of (step 2) implies that

$$\tilde{e}_i (L(\lambda) \otimes L(\infty))_\mu \subset v^{-n} (L(\lambda) \otimes L(\infty))_{\mu+\alpha_i}.$$

Therefore, $\phi_\lambda(\tilde{f}_{i_1} \cdots \tilde{f}_{i_N} 1) \equiv u \otimes 1 \pmod{vL(\lambda) \otimes L(\infty)}$ implies that

$$\phi_\lambda(\tilde{e}_i \tilde{f}_{i_1} \cdots \tilde{f}_{i_N} 1) - \tilde{e}_i(u \otimes 1) \in v^{-n+1} (L(\lambda) \otimes L(\infty))_{\mu+\alpha_i}. \tag{9.7}$$

Adding (9.6) and (9.7), we get

$$\phi_\lambda(\tilde{e}_i \tilde{f}_{i_1} \cdots \tilde{f}_{i_N} 1) - (\tilde{e}_i u) \otimes 1 \in v^{-n+1} L(\lambda) \otimes L(\infty). \tag{9.8}$$

Since $\phi_\lambda(\tilde{e}_i \tilde{f}_{i_1} \cdots \tilde{f}_{i_N} 1)$ belongs to $\phi_\lambda(L(\infty)_{\mu+\alpha_i}) \subset L(\lambda) \otimes L(\infty)$ by the fifth inclusion of (step 2), we conclude that $(\tilde{e}_i u) \otimes 1 \in v^{-n+1} L(\lambda) \otimes L(\infty)$ if $n \geq 1$. Looking at the first component, we also have $\tilde{e}_i u \in v^{-n+1} L(\lambda)_{\mu+\alpha_i}$. Therefore, $\tilde{e}_i L(\lambda)_\mu \subset v^{-n} L(\lambda)_{\mu+\alpha_i}$ implies that $\tilde{e}_i L(\lambda)_\mu \subset v^{-n+1} L(\lambda)_{\mu+\alpha_i}$ whenever $n \geq 1$. As a result, we obtain $\tilde{e}_i L(\lambda)_\mu \subset L(\lambda)_{\mu+\alpha_i}$.

We set $n = 0$ and reread the argument in the induction loop for $n = 0$. Then the argument is still valid until (9.8), so we conclude that if $u = \tilde{f}_{i_1} \cdots \tilde{f}_{i_N} v_\lambda \in L(\lambda)_\mu$ and $u \not\equiv 0 \pmod{vL(\lambda)}$, then

$$\tilde{e}_i(u \otimes 1) \equiv (\tilde{e}_i u) \otimes 1 \pmod{vL(\lambda) \otimes L(\infty)},$$
$$\tilde{e}_i (L(\lambda) \otimes L(\infty))_\mu \subset (L(\lambda) \otimes L(\infty))_{\mu+\alpha_i},$$
$$\phi_\lambda(\tilde{e}_i \tilde{f}_{i_1} \cdots \tilde{f}_{i_N} 1) \equiv \tilde{e}_i(u \otimes 1) \pmod{vL(\lambda) \otimes L(\infty)}.$$

Hence the result.

(step 9) *We have* $L(\lambda)_\mu = \bigoplus_{n \geq 0} f_i^{(n)} (\, L(\lambda)_{\mu+n\alpha_i} \cap \operatorname{Ker} e_i \,)$.

We follow the proof of (step 1). The induction argument does not break down by virtue of (step 8). Namely, by taking the $L(\lambda) \otimes 1$ component of (9.5), we can prove that if $u \in L(\lambda)_\mu$ then $\tilde{e}_i u \in L(\lambda)_{\mu+\alpha_i}$.

(step 10) *If* $b \in B_{\mu-\lambda}$ *and* $bv_\lambda \neq 0$ *then*

$$\tilde{f}_i(bv_\lambda) \equiv (\tilde{f}_i b) v_\lambda \pmod{vL(\lambda)}.$$

Write $b \equiv \tilde{f}_{i_1} \cdots \tilde{f}_{i_N} 1 \pmod{vL(\infty)}$. Then (step 5) implies that

$$bv_\lambda \equiv \tilde{f}_{i_1} \cdots \tilde{f}_{i_N} v_\lambda \not\equiv 0 \pmod{vL(\lambda)},$$

and (step 7) implies that

$$\phi_\lambda(\tilde{f}_{i_1} \cdots \tilde{f}_{i_N} 1) \equiv (\tilde{f}_{i_1} \cdots \tilde{f}_{i_N} v_\lambda) \otimes 1 \pmod{vL(\lambda) \otimes L(\infty)}.$$

Note that Lemma 9.3(1) implies that $\tilde{f}_i \left(L(\lambda) \otimes L(\infty)\right)_\mu \subset \left(L(\lambda) \otimes L(\infty)\right)_{\mu-\alpha_i}$ since (step 9) is already confirmed. Thus

$$\tilde{f}_i(L(\lambda)_\mu \otimes 1) \subset (L(\lambda) \otimes L(\infty))_{\mu-\alpha_i}.$$

Together with the fourth inclusion of (step 2), this implies that

$$\tilde{f}_i \left(L(\lambda) \otimes L(\infty)\right)_\mu \subset \left(L(\lambda) \otimes L(\infty)\right)_{\mu-\alpha_i}. \tag{9.9}$$

Thus we obtain

$$\phi_\lambda(\tilde{f}_i \tilde{f}_{i_1} \cdots \tilde{f}_{i_N} 1) \equiv \tilde{f}_i \left((\tilde{f}_{i_1} \cdots \tilde{f}_{i_N} v_\lambda) \otimes 1 \right) \pmod{vL(\lambda) \otimes L(\infty)}.$$

Since (9.9) also implies that the sixth inclusion of (step 2) holds for $\mu - \alpha_i$, and that $\tilde{f}_i b - \tilde{f}_i \tilde{f}_{i_1} \cdots \tilde{f}_{i_N} 1 \in vL(\infty)$, we get

$$\phi_\lambda(\tilde{f}_i b) \equiv \phi_\lambda(\tilde{f}_i \tilde{f}_{i_1} \cdots \tilde{f}_{i_N} 1) \pmod{vL(\lambda) \otimes L(\infty)}.$$

Thus we have

$$\phi_\lambda(\tilde{f}_i b) \equiv \tilde{f}_i \left((\tilde{f}_{i_1} \cdots \tilde{f}_{i_N} v_\lambda) \otimes 1 \right) \pmod{vL(\lambda) \otimes L(\infty)}. \tag{9.10}$$

Note that (step 3) also works for μ since (step 9) is confirmed. Hence, by taking the $L(\lambda) \otimes 1$ components of (9.10) we get

$$(\tilde{f}_i b) v_\lambda \equiv \tilde{f}_i \tilde{f}_{i_1} \cdots \tilde{f}_{i_N} v_\lambda \pmod{vL(\lambda)}. \tag{9.11}$$

As $bv_\lambda \equiv \tilde{f}_{i_1} \cdots \tilde{f}_{i_N} v_\lambda \pmod{vL(\lambda)}$, we may replace the right hand side of (9.11) with $\tilde{f}_i(bv_\lambda)$. Hence we have the result.

(step 11) *If $b \in B_{\mu-\lambda}$ and $bv_\lambda \neq 0$ then*

$$\tilde{e}_i(bv_\lambda) \equiv (\tilde{e}_i b) v_\lambda \pmod{vL(\lambda)}.$$

Write $b \equiv \tilde{f}_{i_1} \cdots \tilde{f}_{i_N} 1 \pmod{vL(\infty)}$. Then the fifth inclusion of (step 2) and (step 5) (step 8) imply that

$$\phi_\lambda(\tilde{e}_i b) \equiv (\tilde{e}_i(bv_\lambda)) \otimes 1 \pmod{vL(\lambda) \otimes L(\infty)}.$$

Looking at the $L(\lambda) \otimes 1$ component of both sides gives the result. □

We have come to the second main theorem. (See [**Kashiwara**, Theorem 2] and [**Lusztig**, Theorem 20.1.4].)

THEOREM 9.6. **(The second main theorem of Kashiwara and Lusztig)** *Let $(L(\lambda), B(\lambda))$ be as in Theorem 9.5. Then we have the following.*

(1) *$(L(\lambda), B(\lambda))$ is a crystal basis of $V(\lambda)$.*
(2) *Any crystal basis of $V(\lambda)$ coincides with $(L(\lambda), B(\lambda))$ up to a scalar multiple.*

PROOF. (1) By Theorem 9.5(1), we have $L(\lambda) = L(\infty)v_\lambda$. Then Proposition 9.1(2) implies that $L(\lambda)$ is a full rank R-lattice of $V(\lambda)$, and that $\{bv_\lambda | bv_\lambda \neq 0\}$ is an R-free basis of $L(\lambda)$. Further, Theorem 9.5(1) says that

$$B(\lambda) = \{bv_\lambda \pmod{vL(\lambda)} \mid bv_\lambda \neq 0\},$$

which implies that $B(\lambda)$ is a $\mathbb{Q}$-basis of $L(\lambda)/vL(\lambda)$. We know that both $L(\lambda)$ and $B(\lambda)$ have a weight space decomposition, and Theorem 9.5(2) implies that both $L(\lambda)$ and $B(\lambda) \cup \{0\}$ are stable under the Kashiwara operators $\tilde{e}_i$ and $\tilde{f}_i$.

To show the remaining properties of the crystal basis, take an element b of the canonical basis B.

If $\tilde{e}_i(bv_\lambda) \not\equiv 0 \pmod{vL(\lambda)}$ then Lemma 8.3(1)(2) and Theorem 9.5(2) imply that there exists $b' \in B$ such that $\tilde{e}_i b \equiv b' \pmod{vL(\infty)}, b'v_\lambda \neq 0$. One property of the crystal basis of U_v^- is that $b \equiv \tilde{f}_i \tilde{e}_i b \pmod{vL(\infty)}$. Hence Theorem 9.5(2) implies that

$$\tilde{f}_i \tilde{e}_i(bv_\lambda) \equiv \tilde{f}_i(b'v_\lambda) \equiv (\tilde{f}_i b')v_\lambda \equiv bv_\lambda \pmod{vL(\lambda)}.$$

If $\tilde{f}_i(bv_\lambda) \not\equiv 0$, then there exists $b' \in B$ such that $\tilde{f}_i b \equiv b' \pmod{vL(\infty)}$ and $b'v_\lambda \neq 0$. Then the property $b \equiv \tilde{e}_i\tilde{f}_i b \pmod{vL(\infty)}$ and Theorem 9.5(2) imply that

$$\tilde{e}_i\tilde{f}_i(bv_\lambda) \equiv \tilde{e}_i(b'v_\lambda) \equiv (\tilde{e}_i b')v_\lambda \equiv bv_\lambda \pmod{vL(\lambda)}.$$

We have proved that $(L(\lambda), B(\lambda))$ is a crystal basis of $V(\lambda)$.
(2) Let (L, B) be a crystal basis of $V(\lambda)$. Multiplying by a suitable scalar, we may assume that

$$L_\lambda = Rv_\lambda,\ v_\lambda \pmod{vL} \in B_\lambda.$$

Then we may prove $(L, B) = (L(\lambda), B(\lambda))$ by the following steps.

(step 1) *For any $b \in B(\lambda) \setminus \{v_\lambda \pmod{vL(\lambda)}\}$, there exists an index i such that $\tilde{e}_i b \neq 0$.*

Write $b = \tilde{f}_{i_1} \cdots \tilde{f}_{i_N} v_\lambda \pmod{vL(\lambda)}$ and define

$$b' = \tilde{f}_{i_2} \cdots \tilde{f}_{i_N} v_\lambda \pmod{vL(\lambda)}.$$

Then $b' \neq 0$ implies $b' \in B(\lambda)$. By (1) and a property of the crystal basis, $\tilde{f}_{i_1} b' = b$ implies that $\tilde{e}_{i_1} b = b' \neq 0$.

(step 2) *Let $\mu \neq \lambda$, and assume that $u \in L(\lambda)_\mu$ satisfies*

$$\tilde{e}_i u \equiv 0 \pmod{vL(\lambda)}$$

for $0 \leq i \leq r-1$. Then $u \equiv 0 \pmod{vL(\lambda)}$.

Write $u \pmod{vL(\lambda)} = \sum_{b \in B(\lambda)} c_b b$ ($c_b \in \mathbb{Q}$). We choose b and fix it for a while. By (step 1), there exists an index i with $\tilde{e}_i b \neq 0$. Then we have

$$\sum_{b' \in B(\lambda)} c_{b'} \tilde{e}_i b' = 0$$

by assumption. Since $\{b' \in B(\lambda) \mid \tilde{e}_i b' \neq 0\}$ is a set of linearly independent elements in $L(\lambda)/vL(\lambda)$, and b belongs to this set, we have $c_b = 0$. Thus, we have proved that $u \equiv 0 \pmod{vL(\lambda)}$.

(step 3) *Let $\mu \neq \lambda$, and assume that $u \in V(\lambda)_\mu$ satisfies $\tilde{e}_i u \in L(\lambda)$ ($0 \leq i \leq r-1$). Then we have $u \in L(\lambda)$.*

It is enough to show that if $u \in v^{-n}L(\lambda)_\mu$ ($n \geq 1$), then $u \in v^{-n+1}L(\lambda)_\mu$. Applying (step 2) to $v^n u$, we get $v^n u \in vL(\lambda)$. Hence the result.

(step 4) *We have $L = L(\lambda)$.*

The inclusion $L(\lambda) \subset L$ is obvious. We prove $L_\mu \subset L(\lambda)_\mu$ by induction on $\mathrm{ht}(\lambda - \mu)$. Let $u \in L_\mu$. Now $\tilde{e}_i L \subset L$, so by the induction hypothesis we have

$$\tilde{e}_i u \in L_{\mu+\alpha_i} \subset L(\lambda)_{\mu+\alpha_i}$$

for $0 \leq i \leq r-1$. Hence (step 3) implies that $u \in L(\lambda)_\mu$.

(step 5) *We have $B = B(\lambda)$.*

Recall that $\tilde{f}_i b \neq 0$ implies $\tilde{f}_i b \in B$ if $b \in B$. Using this, we may prove $B(\lambda) \subset B$ by induction on $\mathrm{ht}(\lambda - \mu)$. Since both $B(\lambda)$ and B are bases of L/vL, we have $B = B(\lambda)$. □

In these notes, we only consider irreducible integrable highest weight modules. The existence and uniqueness of a crystal basis may be proved for integrable modules which belong to "category $\mathcal{O}$". In other words, if an integrable U_v-module is a direct sum of irreducible integrable highest weight U_v-modules then there exists a unique crystal basis up to isomorphism. See [**Kashiwara**, Theorem 3].

Before closing this section, we will prove two more theorems. The third main theorem is obvious if one follows Lusztig's approach as we have done. If one follows Kashiwara's approach, this is proved simultaneously with the other main theorems in Kashiwara's "grand loop". See [**Kashiwara**, Theorem 5] for the Kashiwara's approach.

DEFINITION 9.7. Let $V(\lambda)$ be an integrable highest weight U_v-module. Since $V(\lambda) \simeq U_v^- / \sum U_v^- f_i^{1+\lambda(h_i)}$ as a U_v^--module by Proposition 9.1(1), the bar involution of U_v^- induces a $\mathbb{Q}$-linear automorphism $uv_\lambda \mapsto \bar{u}v_\lambda$ of $V(\lambda)$. We call this the **bar involution** of $V(\lambda)$.

THEOREM 9.8. **(Characterization of the canonical basis of $V(\lambda)$)**
Let $V(\lambda)$ be an integrable highest weight U_v-module and $(L(\lambda), B(\lambda))$ the crystal basis defined above. Then there exists a unique basis B of $V(\lambda)$ with the following properties.

(1) $B \subset L(\lambda)$,
(2) $\bar{b} = b$ $(b \in B)$,
(3) $B(\lambda) = \{b \pmod{vL(\lambda)}\}_{b \in B}$.

PROOF. The uniqueness is easy to prove, and the existence follows from Theorem 7.3(2) and Theorem 9.5(1). □

THEOREM 9.9. **(The third main theorem of Kashiwara and Lusztig)**
Let (L, B) be a crystal basis of U_v^-. Then $(Lv_\lambda, \{bv_\lambda \mid b \in B\} \setminus \{0\})$ is a crystal basis of $V(\lambda)$.

PROOF. This follows from Theorem 8.5(2), Theorem 9.5(1), and Theorem 9.6(2). □

CHAPTER 10

The Hayashi realization

10.1. Partitions and the Hayashi realization

We have proved in the previous chapter that each irreducible integrable highest weight U_v-module has a crystal basis which is unique up to a scalar multiple. But its construction is not at all combinatorial. Hence a natural question arises: "can we construct the crystal basis by combinatorial methods ?" For the quantum algebra of type $A_{r-1}^{(1)}$ this is in fact possible and there are two ways to do so. We are now in the heart of these lectures. We shall explain this theory in this and the next chapter. From now on, the theory we develop is specific to type $A_{r-1}^{(1)}$.

In this chapter we construct the representation $V(\Lambda_0)$ of U_v combinatorially. Its original construction was obtained by Hayashi, and reformulated into combinatorial language by Misra and Miwa; see [**mp-H**] and [**mp-MM**]. The proof given here is new.

DEFINITION 10.1. A **partition** is a sequence of non-increasing natural numbers $\lambda = (\lambda_1, \dots, \lambda_l)$. The corresponding **Young diagram** is a collection of rows of square boxes which are left justified and with λ_i boxes in the ith row $(1 \le i \le l)$. We do not distinguish between a partition and the corresponding Young diagram.

A box in a Young diagram is called a **node**. If x is a node of λ we write $x \in \lambda$. We denote the row number and the column number of the node x by $\mathrm{row}(x)$ and $\mathrm{col}(x)$ respectively. We denote the number of rows in λ by $l(\lambda)$ and the number of nodes in λ by $|\lambda|$. If $|\lambda| = n$, we write $\lambda \vdash n$.

Assume that we are given a positive integer r. Then the **r-residue** of a node x is defined by

$$\mathrm{res}(x) = -\mathrm{row}(x) + \mathrm{col}(x) \pmod r \in \mathbb{Z}/r\mathbb{Z}.$$

EXAMPLE 10.2. The following diagram is the Young diagram $\lambda = (53211)$, which has $l(\lambda) = 5$ rows. The 3-residue of the node x in the diagram with $\mathrm{row}(x) = 4, \mathrm{col}(x) = 1$ is $\mathrm{res}(x) = 0$.

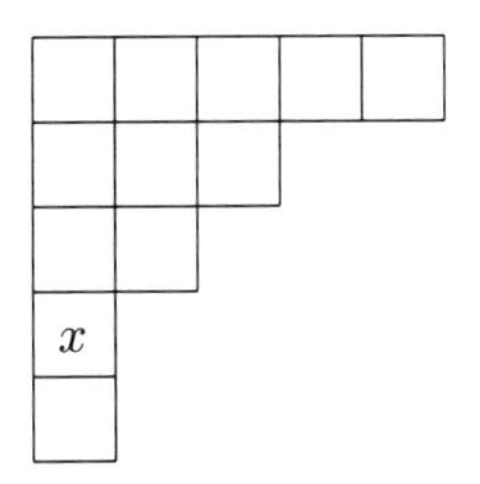

DEFINITION 10.3. Let λ be a Young diagram. If we get a Young diagram μ by adding a node to λ, we denote this node by μ/λ. Such a node is called an addable node of λ. If an addable node of λ has residue i, we call it an **addable i-node** of λ. We denote the set of addable i-nodes of λ by $A_i(\lambda)$.

Similarly, if we get a Young diagram μ by deleting a node from λ, we denote this node by λ/μ. Such a node is called a removable node of λ. If a removable node of λ has r-residue i, we call it a **removable i-node** of λ. We denote the set of removable i-nodes of λ by $R_i(\lambda)$. As long as λ is fixed, we may identify $A_i(\lambda), R_i(\lambda)$ with subsets of $\mathbb{N}$ by sending $x \in A_i(\lambda)$ or $R_i(\lambda)$ to $\mathrm{row}(x)$.

EXAMPLE 10.4. If we fill in the nodes of the Young diagram $\lambda = (53211)$ with 3-residues, we have the diagram below. We have $A_2(\lambda) = \{y, z\}$ where y and z are the nodes displayed in the diagram.

0	1	2	0	1	z
2	0	1	y		
1	2				
0					
2					

DEFINITION 10.5. Let $\mathcal{F}$ be the K-vector space whose basis is the set of all Young diagrams. For a Young diagram λ, and a pair of Young diagrams λ and μ such that $|\lambda| - |\mu| = \pm 1$, we define $N_i(\lambda), W_i(\lambda)$, $N_i^a(\lambda/\mu)$ and $N_i^b(\mu/\lambda)$ $(0 \le i < r)$ as follows:

$$N_i(\lambda) = |A_i(\lambda)| - |R_i(\lambda)|, \quad W_i(\lambda) = |\{\, x \in \lambda \mid \mathrm{res}(x) = i \}|,$$

$$N_i^a(\lambda/\mu) = |\{\, x \in A_i(\lambda) \mid \mathrm{row}(x) < \mathrm{row}(\lambda/\mu)\}| \\ - |\{\, x \in R_i(\lambda) \mid \mathrm{row}(x) < \mathrm{row}(\lambda/\mu)\}|,$$

$$N_i^b(\mu/\lambda) = |\{\, x \in A_i(\lambda) \mid \mathrm{row}(x) > \mathrm{row}(\mu/\lambda)\}| \\ - |\{\, x \in R_i(\lambda) \mid \mathrm{row}(x) > \mathrm{row}(\mu/\lambda)\}|.$$

THEOREM 10.6 (The Hayashi realization). *Let $\mathcal{F}$, $N_i(\lambda), W_i(\lambda)$, $N_i^a(\lambda/\mu)$ and $N_i^b(\mu/\lambda)$ be as above. Then $\mathcal{F}$ becomes an integrable U_v-module via the following action.*

$$e_i\lambda = \sum_{\mathrm{res}(\lambda/\mu)\equiv i} v^{-N_i^a(\lambda/\mu)}\mu, \quad f_i\lambda = \sum_{\mathrm{res}(\mu/\lambda)\equiv i} v^{N_i^b(\mu/\lambda)}\mu,$$

$$t_i\lambda = v^{N_i(\lambda)}\lambda, \quad v^d\lambda = v^{-W_0(\lambda)}\lambda.$$

The sum in the definition of e_i and f_i above runs over all μ that satisfy the given condition on the r-residue, and $\equiv$ means "modulo r".

The U_v-submodule generated by the empty Young diagram is isomorphic to $V(\Lambda_0)$.

PROOF. We prove that $\mathcal{F}$ becomes a U_v-module. Once this is proved, the integrability follows directly from the definition of the action and Lemma 6.9(3). The computation of the highest weight proves that $\emptyset$ generates an irreducible U_v-submodule which is isomorphic to $V(\Lambda_0)$.

We check the defining relations of U_v one by one.

(1) *Any two of t_i $(0 \le i \le r-1)$ and v^d commute.*

This follows from the definition of the action of t_i and v^d.

(2) $t_i e_j t_i^{-1} = v^{\alpha_j(h_i)} e_j$. *(The proof that $t_i f_j t_i^{-1} = v^{-\alpha_j(h_i)} f_j$ is similar.)*

Let μ be a Young diagram which appears in $e_j\lambda$. We have to show that

$$N_i(\mu) - N_i(\lambda) = \alpha_j(h_i). \tag{10.1}$$

Assume that μ is obtained from λ by deleting a node in the k_0th row. So we have

$$\lambda_{k_0-1} \ge \lambda_{k_0} > \lambda_{k_0+1}, \quad -k_0 + \lambda_{k_0} \equiv j.$$

We identify $A_i(\lambda)$ and $A_i(\mu)$ with subsets of $\mathbb{N}$ by using the row numbers of the nodes. Then we may describe these sets as follows:

$$\begin{aligned} A_i(\lambda) &= \{k \mid \lambda_{k-1} > \lambda_k,\ -k + \lambda_k + 1 \equiv i\}, \\ A_i(\mu) &= \{k \mid \mu_{k-1} > \mu_k,\ -k + \mu_k + 1 \equiv i\}. \end{aligned}$$

For these formulas we set $\lambda_0 = \mu_0 = \infty$ and $\lambda_k = 0$ for $k > l(\lambda)$, $\mu_k = 0$ for $k > l(\mu)$.

We obviously have that if $k \ne k_0, k_0+1$, then $k \in A_i(\lambda) \Leftrightarrow k \in A_i(\mu)$. Otherwise, we have the following formulas.

$$\begin{cases} k_0 \in A_i(\lambda) & \Leftrightarrow \lambda_{k_0-1} > \lambda_{k_0},\ j \equiv i-1, \\ k_0 \in A_i(\mu) & \Leftrightarrow j \equiv i, \end{cases}$$

$$\begin{cases} k_0 + 1 \in A_i(\lambda) & \Leftrightarrow -k_0 + \lambda_{k_0+1} \equiv i, \\ k_0 + 1 \in A_i(\mu) & \Leftrightarrow \lambda_{k_0} > \lambda_{k_0+1} + 1,\ -k_0 + \lambda_{k_0+1} \equiv i. \end{cases}$$

Similarly, if we identify $R_i(\lambda)$ and $R_i(\mu)$ with subsets of $\mathbb{N}$, we may describe them by

$$\begin{aligned} R_i(\lambda) &= \{k \mid \lambda_k > \lambda_{k+1},\ -k + \lambda_k \equiv i\}, \\ R_i(\mu) &= \{k \mid \mu_k > \mu_{k+1},\ -k + \mu_k \equiv i\}. \end{aligned}$$

Thus, if $k \ne k_0 - 1, k_0$ then $k \in R_i(\lambda) \Leftrightarrow k \in R_i(\mu)$; otherwise,

$$\begin{cases} k_0 \in R_i(\lambda) & \Leftrightarrow j \equiv i, \\ k_0 \in R_i(\mu) & \Leftrightarrow \lambda_{k_0} > \lambda_{k_0+1} + 1,\ j \equiv i+1, \end{cases}$$

$$\begin{cases} k_0 - 1 \in R_i(\lambda) & \Leftrightarrow \lambda_{k_0-1} > \lambda_{k_0},\ -k_0 + \lambda_{k_0-1} \equiv i-1, \\ k_0 - 1 \in R_i(\mu) & \Leftrightarrow -k_0 + \lambda_{k_0-1} \equiv i-1. \end{cases}$$

To prove (10.1), we consider four cases.

(case 1) $\lambda_{k_0-1} > \lambda_{k_0} > \lambda_{k_0+1} + 1$

We have the following formulas.

$$|A_i(\mu)| - |A_i(\lambda)| = \begin{cases} 1 & (j \equiv i), \\ -1 & (j \equiv i-1), \\ 0 & (otherwise). \end{cases}$$

$$|R_i(\mu)| - |R_i(\lambda)| = \begin{cases} -1 & (j \equiv i), \\ 1 & (j \equiv i+1), \\ 0 & (otherwise). \end{cases}$$

Thus, if $r \geq 3$ then

$$N_i(\mu) - N_i(\lambda) = \begin{cases} 2 & (j \equiv i), \\ -1 & (j \equiv i \pm 1), \\ 0 & (otherwise). \end{cases}$$

and if $r = 2$ then

$$N_i(\mu) - N_i(\lambda) = \begin{cases} 2 & (j \equiv i), \\ -2 & (j \equiv i+1). \end{cases}$$

Hence the result follows.

(case 2) $\lambda_{k_0-1} = \lambda_{k_0} > \lambda_{k_0+1} + 1$

If $\lambda_{k_0-1} = \lambda_{k_0}$ then we have $-k_0 + \lambda_{k_0-1} \equiv i-1 \Leftrightarrow j \equiv i-1$. Thus, the formulas differ from (case 1) only when $j = i-1$. In this case, both $|A_i(\mu)| - |A_i(\lambda)|$ and $|R_i(\mu)| - |R_i(\lambda)|$ increase by 1. Hence the result.

(case 3) $\lambda_{k_0-1} > \lambda_{k_0} = \lambda_{k_0+1} + 1$

If $\lambda_{k_0} = \lambda_{k_0+1} + 1$ then we have $-k_0 + \lambda_{k_0+1} \equiv i \Leftrightarrow j \equiv i+1$. Thus, the formulas differ from (case 1) only when $j = i+1$. In this case, both $|A_i(\mu)| - |A_i(\lambda)|$ and $|R_i(\mu)| - |R_i(\lambda)|$ decrease by 1. Hence the result.

(case 4) $\lambda_{k_0-1} = \lambda_{k_0} = \lambda_{k_0+1} + 1$

The proof is the same as (case 2) and (case 3).

(3) $v^d e_j v^{-d} = v^{\delta_{0j}} e_j$. *(The proof that $v^d f_j v^{-d} = v^{-\delta_{0j}} f_j$ is similar.)*

Let μ be a Young diagram which appears in $e_j\lambda$. We have to show that

$$-W_0(\mu) + W_0(\lambda) = \delta_{0j},$$

but this is obvious.

(4) $e_i f_j - f_j e_i = \delta_{ij} \frac{t_i - t_i^{-1}}{v - v^{-1}}$.

Let $\lambda \neq \mu$. If μ appears in $e_i f_j \lambda$ then there exists a Young diagram ν which satisfies the condition (10.2) below. Then the existence of the Young diagram ν' guarantees that μ appears in $f_j e_i \lambda$. Conversely, if μ appears in $f_j e_i \lambda$ then a similar argument shows that μ appears in $e_i f_j \lambda$. Hence, to see the coefficient of $\mu \neq \lambda$ in $(e_i f_j - f_j e_i)\lambda$, it is enough to consider μ with (10.3).

$$\begin{gathered} \nu = \lambda \cup \{z\} = \mu \cup \{y\}, \quad \nu' = \lambda \setminus \{y\} = \mu \setminus \{z\}, \\ r(y) = i, \quad r(z) = j. \end{gathered} \tag{10.2}$$

$$
\begin{array}{ccc}
\lambda & \subset & \nu \\
\cup & & \cup \\
\nu' & \subset & \mu
\end{array}
\tag{10.3}
$$

We shall prove that the coefficients of $\mu \neq \lambda$ in $e_i f_j \lambda$ and $f_j e_i \lambda$ are the same. More explicitly, we prove that

$$N_j^b(\nu/\lambda) - N_i^a(\nu/\mu) = -N_i^a(\lambda/\nu') + N_j^b(\mu/\nu').$$

Let $\mathrm{row}(y) = k_0, \mathrm{row}(z) = k_1$, and let $\lambda_{>k_1}$ and $\nu'_{>k_1}$ be the Young diagrams obtained from λ and ν' by deleting the first to the k_1th rows. We have

$$
\begin{aligned}
N_j^b(\nu/\lambda) - N_j^b(\mu/\nu') = |\{x \in A_j(\lambda) | \mathrm{row}(x) > \mathrm{row}(z)\}| \\
- |\{x \in R_j(\lambda) | \mathrm{row}(x) > \mathrm{row}(z)\}| \\
- |\{x \in A_j(\nu') | \mathrm{row}(x) > \mathrm{row}(z)\}| \\
+ |\{x \in R_j(\nu') | \mathrm{row}(x) > \mathrm{row}(z)\}|.
\end{aligned}
$$

We may consider $N_j^b(\nu/\lambda) - N_j^b(\mu/\nu')$ as $N_j(\lambda_{>k_1}) - N_j(\nu'_{>k_1})$ with the 0th rows $(\lambda_{>k_1})_0 < \infty$ and $(\mu_{>k_1})_0 < \infty$. If $k_0 \leq k_1$ this is 0. If $k_0 > k_1$, then we have $\nu'_{>k_1} \cup \{y\} = \lambda_{>k_1}$, and $\lambda_{>k_1}$ appears in $e_i \nu'_{>k_1}$. Hence we may apply (2). Note that the argument in (2) works not only for $\lambda_0 = \mu_0 = \infty$ but also for finite λ_0 and μ_0. The result is that

$$N_j^b(\nu/\lambda) - N_j^b(\mu/\nu') = \begin{cases} \alpha_i(h_j) & (k_0 > k_1), \\ 0 & (\textit{otherwise}). \end{cases}$$

Similarly, let $\nu_{<k_0}$ and $\lambda_{<k_0}$ be the Young diagrams obtained from ν and λ by deleting the k_0th to the last rows. We have

$$
\begin{aligned}
N_i^a(\nu/\mu) - N_i^a(\lambda/\nu') = |\{x \in A_i(\nu) | \mathrm{row}(x) < \mathrm{row}(y)\}| \\
- |\{x \in R_i(\nu) | \mathrm{row}(x) < \mathrm{row}(y)\}| \\
- |\{x \in A_i(\lambda) | \mathrm{row}(x) < \mathrm{row}(y)\}| \\
+ |\{x \in R_i(\lambda) | \mathrm{row}(x) < \mathrm{row}(y)\}|.
\end{aligned}
$$

We may consider $N_i^a(\nu/\mu) - N_i^a(\lambda/\nu')$ as $N_i(\nu_{<k_0}) - N_i(\lambda_{<k_0})$ with finite 0th rows. If $\lambda_{<k_0} = \nu_{<k_0}$ this is 0, and if $\lambda_{<k_0} \cup \{z\} = \nu_{<k_0}$ we apply (2) as above. The result is

$$N_i^a(\nu/\mu) - N_i^a(\lambda/\nu') = \begin{cases} \alpha_j(h_i) & (k_0 > k_1), \\ 0 & (\textit{otherwise}). \end{cases}$$

Since $\alpha_i(h_j) = \alpha_j(h_i)$, we obtain

$$N_j^b(\nu/\lambda) - N_j^b(\mu/\nu') = N_i^a(\nu/\mu) - N_i^a(\lambda/\nu'),$$

from which the result follows.

A consequence of this result is that $(e_i f_j - f_j e_i)\lambda = 0$ if $i \neq j$. If $i = j$, then $(e_i f_j - f_j e_i)\lambda$ is a multiple of λ. To determine this multiple, we define $a_i(k), b_i(k)$ $(k \in \mathbb{N})$ as follows.

$$
\begin{aligned}
a_i(k) = |\{\, k' < k \mid \lambda_{k'-1} > \lambda_{k'}, -k' + \lambda_{k'} + 1 \equiv i\}| \\
- |\{\, k' < k \mid \lambda_{k'} > \lambda_{k'+1}, -k' + \lambda_{k'} \equiv i\}|,
\end{aligned}
$$

$$b_i(k) = |\{\, k' > k \mid \lambda_{k'-1} > \lambda_{k'}, -k' + \lambda_{k'} + 1 \equiv i\}| \\ - |\{\, k' > k \mid \lambda_{k'} > \lambda_{k'+1}, -k' + \lambda_{k'} \equiv i\}|.$$

Let A_i, R_i and E_i be subsets of $\mathbb{N}$ defined by

$$\begin{aligned} A_i &= \{\, k \mid \lambda_{k-1} > \lambda_k,\ -k + \lambda_k + 1 \equiv i\}, \\ R_i &= \{\, k \mid \lambda_k > \lambda_{k+1},\ -k + \lambda_k \equiv i\}, \\ E_i &= \{k \mid k \notin A_i \cup R_i\}. \end{aligned}$$

Then we have a partition $\mathbb{N} = A_i \sqcup R_i \sqcup E_i$ and the following hold.

$$(e_i f_i - f_i e_i)\lambda = \left(\sum_{k\in A_i} v^{-a_i(k)+b_i(k)} - \sum_{k\in R_i} v^{-a_i(k)+b_i(k)} \right) \lambda,$$

$$a_i(k) - a_i(k+1) = \begin{cases} -1 & (k \in A_i) \\ 1 & (k \in R_i) \\ 0 & (k \in E_i) \end{cases}, \tag{10.4}$$

and

$$b_i(k) - b_i(k+1) = \begin{cases} 1 & (k+1 \in A_i) \\ -1 & (k+1 \in R_i) \\ 0 & (k+1 \in E_i) \end{cases}. \tag{10.5}$$

We have to show that

$$\sum_{k\in A_i} v^{-a_i(k)+b_i(k)} - \sum_{k\in R_i} v^{-a_i(k)+b_i(k)} = \frac{v^{N_i(\lambda)} - v^{-N_i(\lambda)}}{v - v^{-1}},$$

which is equivalent to

$$\sum_{k\in A_i} v^{-a_i(k)+b_i(k)+1} + \sum_{k\in R_i} v^{-a_i(k)+b_i(k)-1} + v^{-N_i(\lambda)} \\ = v^{N_i(\lambda)} + \sum_{k\in A_i} v^{-a_i(k)+b_i(k)-1} + \sum_{k\in R_i} v^{-a_i(k)+b_i(k)+1}.$$

To prove this, we define $\Delta_i(k)$ and $\Delta_i'(k)$ as follows.

$$\Delta_i(k) = \begin{cases} -a_i(k) + b_i(k) + 1 & (k \in A_i) \\ -a_i(k) + b_i(k) - 1 & (k \in R_i) \\ -a_i(k) + b_i(k) & (k \in E_i) \end{cases},$$

$$\Delta_i'(k) = \begin{cases} -a_i(k) + b_i(k) - 1 & (k \in A_i) \\ -a_i(k) + b_i(k) + 1 & (k \in R_i) \\ -a_i(k) + b_i(k) & (k \in E_i) \end{cases}.$$

Then the formulas (10.4) and (10.5) for $a_i(k) - a_i(k+1)$, $b_i(k) - b_i(k+1)$ imply that

$$\Delta_i(k) - \Delta_i(k+1) = \begin{cases} 2 & (k \in A_i) \\ -2 & (k \in R_i) \\ 0 & (k \in E_i) \end{cases},$$

$$\Delta'_i(k) - \Delta'_i(k+1) = \begin{cases} 2 & (k+1 \in A_i) \\ -2 & (k+1 \in R_i) \\ 0 & (k+1 \in E_i) \end{cases}.$$

Therefore, the boundary condition

$$\Delta_i(k) = \Delta'_i(k) = -N_i(\lambda) \ (k >> 0)$$

shows that $\Delta_i(k) = \Delta'_i(k-1)$, proving (4).

(5) *The relation between e_i and e_j holds if $r \geq 3$. (The proof of the relation between f_i and f_j is similar.)*

We have to show that

$$e_i^2 e_j - (v + v^{-1}) e_i e_j e_i + e_j e_i^2 = 0 \quad (i - j \equiv \pm 1), \tag{10.6}$$

$$e_i e_j - e_j e_i = 0 \quad (otherwise). \tag{10.7}$$

Let j be the r-residue of the removable node of λ lying on the k_0th row, and let μ be the Young diagram obtained from λ by deleting this node. We study $a_i(k)$ introduced in (4). To do this, first we define $A_i(\lambda)_{<k}$ and $R_i(\lambda)_{<k}$ as follows.

$$A_i(\lambda)_{<k} = \{\, k' < k \mid \lambda_{k'-1} > \lambda_{k'},\ -k' + \lambda_{k'} + 1 \equiv i \},$$
$$R_i(\lambda)_{<k} = \{\, k' < k \mid \lambda_{k'} > \lambda_{k'+1},\ -k' + \lambda_{k'} \equiv i \}.$$

We apply the similar argument as in (2) to $\lambda_{<k}$ and $\mu_{<k}$, and get the following result.

Assertion 1 *The values of $|A_i(\mu)_{<k}| - |A_i(\lambda)_{<k}|$ are as follows.*

- If $\lambda_{k_0-1} > \lambda_{k_0} > \lambda_{k_0+1} + 1$ then

$$|A_i(\mu)_{<k}| - |A_i(\lambda)_{<k}| = \begin{cases} 1 & (k \geq k_0 + 1,\ j \equiv i) \\ -1 & (k \geq k_0 + 1,\ j \equiv i - 1) \\ 0 & (otherwise) \end{cases}.$$

- If $\lambda_{k_0-1} = \lambda_{k_0} > \lambda_{k_0+1} + 1$ then

$$|A_i(\mu)_{<k}| - |A_i(\lambda)_{<k}| = \begin{cases} 1 & (k \geq k_0 + 1,\ j \equiv i) \\ 0 & (otherwise) \end{cases}.$$

- If $\lambda_{k_0-1} > \lambda_{k_0} = \lambda_{k_0+1} + 1$ then

$$|A_i(\mu)_{<k}| - |A_i(\lambda)_{<k}| = \begin{cases} 1 & (k \geq k_0 + 1,\ j \equiv i) \\ -1 & (k \geq k_0 + 1,\ j \equiv i - 1) \\ -1 & (k \geq k_0 + 2,\ j \equiv i + 1) \\ 0 & (otherwise) \end{cases}.$$

- If $\lambda_{k_0-1} = \lambda_{k_0} = \lambda_{k_0+1} + 1$ then

$$|A_i(\mu)_{<k}| - |A_i(\lambda)_{<k}| = \begin{cases} 1 & (k \geq k_0 + 1,\ j \equiv i) \\ -1 & (k \geq k_0 + 2,\ j \equiv i + 1) \\ 0 & (otherwise) \end{cases}.$$

Assertion 2 *The values of $|R_i(\mu)_{<k}| - |R_i(\lambda)_{<k}|$ are as follows.*

- If $\lambda_{k_0-1} > \lambda_{k_0} > \lambda_{k_0+1} + 1$ then

$$|R_i(\mu)_{<k}| - |R_i(\lambda)_{<k}| = \begin{cases} -1 & (k \geq k_0 + 1,\ j \equiv i) \\ 1 & (k \geq k_0 + 1,\ j \equiv i + 1) \\ 0 & (otherwise) \end{cases}.$$

- If $\lambda_{k_0-1} = \lambda_{k_0} > \lambda_{k_0+1} + 1$ then

$$|R_i(\mu)_{<k}| - |R_i(\lambda)_{<k}| = \begin{cases} -1 & (k \geq k_0 + 1,\ j \equiv i) \\ 1 & (k \geq k_0 + 1,\ j \equiv i + 1) \\ 1 & (k \geq k_0,\ j \equiv i - 1) \\ 0 & (otherwise) \end{cases}.$$

- If $\lambda_{k_0-1} > \lambda_{k_0} = \lambda_{k_0+1} + 1$ then

$$|R_i(\mu)_{<k}| - |R_i(\lambda)_{<k}| = \begin{cases} -1 & (k \geq k_0 + 1,\ j \equiv i) \\ 0 & (otherwise) \end{cases}.$$

- If $\lambda_{k_0-1} = \lambda_{k_0} = \lambda_{k_0+1} + 1$ then

$$|R_i(\mu)_{<k}| - |R_i(\lambda)_{<k}| = \begin{cases} -1 & (k \geq k_0 + 1,\ j \equiv i) \\ 1 & (k \geq k_0,\ j \equiv i - 1) \\ 0 & (otherwise) \end{cases}.$$

Therefore, the difference $\Delta a_i(k)$ defined by

$$\Delta a_i(k) = (|A_i(\mu)_{<k}| - |R_i(\mu)_{<k}|) - (|A_i(\lambda)_{<k}| - |R_i(\lambda)_{<k}|)$$

is given by the following formulas.

- If $\lambda_{k_0-1} > \lambda_{k_0} > \lambda_{k_0+1} + 1$ then

$$\Delta a_i(k) = \begin{cases} 2 & (k \geq k_0 + 1,\ j \equiv i) \\ -1 & (k \geq k_0 + 1,\ j \equiv i \pm 1) \\ 0 & (otherwise) \end{cases}.$$

The nodes which affect this difference are the nodes pointed to in the picture below.

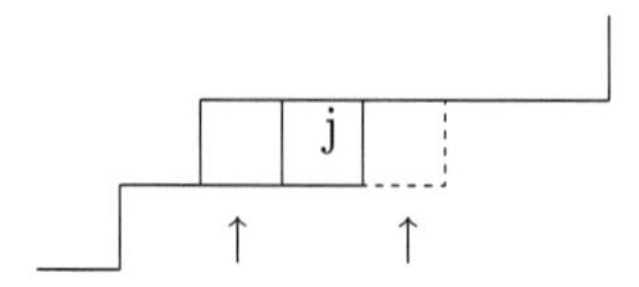

- If $\lambda_{k_0-1} = \lambda_{k_0} > \lambda_{k_0+1} + 1$ then

$$\Delta a_i(k) = \begin{cases} 2 & (k \geq k_0 + 1,\ j \equiv i) \\ -1 & (k \geq k_0 + 1,\ j \equiv i + 1) \\ -1 & (k \geq k_0,\ j \equiv i - 1) \\ 0 & (otherwise) \end{cases}.$$

The nodes which affect this difference are the nodes pointed to in the picture below.

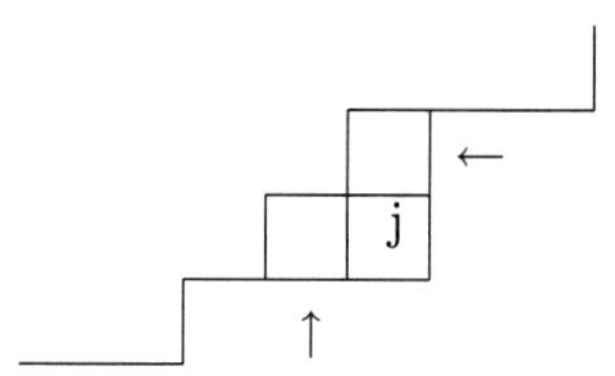

- If $\lambda_{k_0-1} > \lambda_{k_0} = \lambda_{k_0+1} + 1$ then

$$\Delta a_i(k) = \begin{cases} 2 & (k \geq k_0 + 1,\ j \equiv i) \\ -1 & (k \geq k_0 + 2,\ j \equiv i + 1) \\ -1 & (k \geq k_0 + 1,\ j \equiv i - 1) \\ 0 & (otherwise) \end{cases}.$$

The nodes which affect this difference are the nodes pointed to in the picture below.

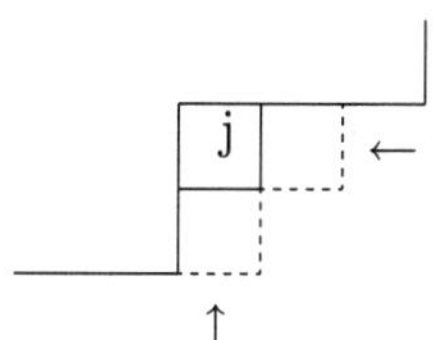

- If $\lambda_{k_0-1} = \lambda_{k_0} = \lambda_{k_0+1} + 1$ then

$$\Delta a_i(k) = \begin{cases} 2 & (k \geq k_0 + 1,\ j \equiv i) \\ -1 & (k \geq k_0 + 2,\ j \equiv i + 1) \\ -1 & (k \geq k_0,\ j \equiv i - 1) \\ 0 & (otherwise) \end{cases}.$$

The nodes which affect this difference are the nodes pointed to in the picture below.

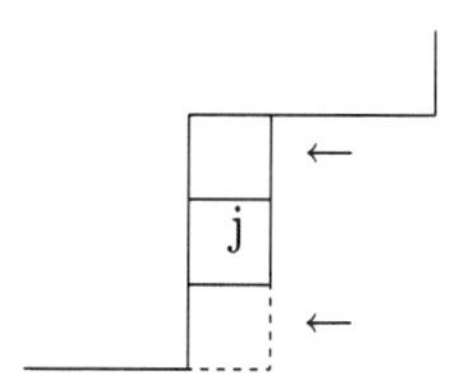

We are now in a position to prove (10.6) and (10.7). We only consider the case where $j \equiv i \pm 1$ (10.6); the other case is easier.

We have to show that the coefficient of μ in $(e_i^2 e_j - (v+v^{-1})e_i e_j e_i + e_j e_i^2)\lambda$ is always 0.

We divide the proof according to the shape of λ/μ. First we assume that λ/μ has three connected components. The following is an example.

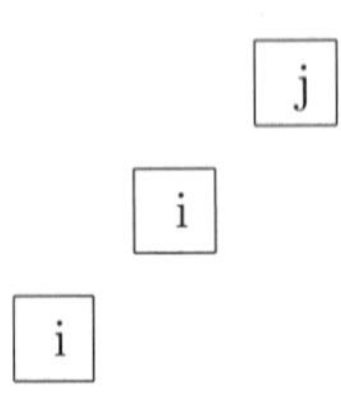

In this case, we denote the row numbers by k_1, k_2, k_3 from top to bottom. Then the coefficient of μ in $e_i^2 e_j \lambda$ is

$$v^{-a_j(k_1)-a_i(k_2)-a_i(k_3)}(1+v^2).$$

To obtain this, first we remove $\boxed{j}$ from the k_1th row. Then the coefficient is multiplied by $v^{-a_j(k_1)}$. Using the formulas for $\Delta a_i(k)$, we have that $a_i(k_2)$ and $a_i(k_3)$ increase by -1, and the new values are $a_i(k_2)-1$ and $a_i(k_3)-1$ respectively. Next we remove $\boxed{i}$. If we remove the node on the k_2th row before the node on the k_3th row, the coefficient is multiplied by $v^{-a_i(k_2)+1}$ and $a_i(k_3)$ increases by 2 to become $a_i(k_3)+1$. Hence, its contribution to the coefficient is $v^{-a_j(k_1)-a_i(k_2)-a_i(k_3)}$. On the other hand, if we remove the node on the k_3th row before the node on the k_2th row, the coefficient is multiplied by $v^{-a_i(k_2)+1}$ and the value of $a_i(k_3)$ remains the same; that is $a_i(k_3)-1$. Hence, its contribution to the coefficient is $v^{-a_j(k_1)-a_i(k_2)-a_i(k_3)+2}$. Thus the coefficient is as desired.

A similar computation shows that the coefficient of μ in $e_i e_j e_i \lambda$ is

$$v^{-a_j(k_1)-a_i(k_2)-a_i(k_3)}(v+v^{-1}),$$

and the coefficient of μ in $e_j e_i^2 \lambda$ is

$$v^{-a_j(k_1)-a_i(k_2)-a_i(k_3)}(1+v^{-2}).$$

Therefore, we have $(e_i^2 e_j - (v+v^{-1})e_i e_j e_i + e_j e_i^2)\lambda = 0$ since

$$(1+v^2)-(v+v^{-1})^2+(1+v^{-2})=0.$$

Similarly, if the r-residues are i, j, i from top to bottom then the relation holds since

$$(v+v^{-1})-2(v+v^{-1})+(v+v^{-1})=0.$$

Finally, if the r-residues are i, i, j then $(1+v^{-2})-(v+v^{-1})^2+(1+v^2)=0$ gives the relation.

If the number of connected components is less than 3 then we divide the cases according to the adjacency of nodes and $j=i+1$ or $j=i-1$, and argue in the same way. We leave this to the reader as an exercise.

(5)' *The defining relations also hold when $r=2$.*

Repeat the proof of (5) but without distinguishing between $j=i+1$ and $j=i-1$. □

10.2. Generalization to the case of multipartitions

Based on Theorem 10.6, we construct another representation of U_v. First we prepare some notation.

DEFINITION 10.7. A **multipartition** is an m-tuple of partitions, which we denote by $\lambda = (\lambda^{(m)}, \dots, \lambda^{(1)})$. If the total number of nodes $|\lambda| := \sum |\lambda^{(k)}|$ is n, we write $\lambda \vdash n$. We identify λ with an m-tuple of Young diagrams, and often call it a (multi-)Young diagram.

Let x, y be two nodes of λ. Then we say that x is **above** y or equivalently y is **below** x, if x is located above y in the following picture.

$$\begin{array}{c}\lambda^{(1)}\\ \cdot\\ \cdot\\ \lambda^{(m)}\end{array}$$

Our goal here is to introduce a U_v-module whose basis is given by multipartitions. To do this, we extend the definition of r-residues as follows.

DEFINITION 10.8. The r**-residue** associated with $(\gamma_m, \dots, \gamma_1) \in (\mathbb{Z}/r\mathbb{Z})^m$ is a map which assigns $\mathrm{res}(x) \in \mathbb{Z}/r\mathbb{Z}$ to each node x where $\mathrm{res}(x)$ is defined by

$$\mathrm{res}(x) = \gamma_{\mathrm{comp}(x)} - \mathrm{row}(x) + \mathrm{col}(x) \pmod r \in \mathbb{Z}/r\mathbb{Z}.$$

Here, by definition x lies on the $\mathrm{row}(x)$th row and the $\mathrm{col}(x)$th column of the component $\lambda^{(\mathrm{comp}(x))}$.

We define $N_i(\lambda)$, $W_i(\lambda)$, $N_i^a(\lambda/\mu)$ and $N_i^b(\mu/\lambda)$ for multipartitions in the same way as before. Namely,

- $N_i(\lambda)$ is the number of addable i-nodes of λ minus the number of removable i-nodes of λ,
- $W_i(\lambda)$ is the number of i-nodes in λ,
- $N_i^a(\lambda/\mu)$ is the number of addable i-nodes of λ lying strictly above λ/μ minus the number of removable i-nodes of λ lying strictly above λ/μ.
- $N_i^b(\mu/\lambda)$ is the number of addable i-nodes of λ lying strictly below μ/λ minus the number of removable i-nodes of λ lying strictly below μ/λ.

DEFINITION 10.9. Let $\mathrm{res}(x)$ be the r-residue associated with $(\gamma_m, \dots, \gamma_1)$. Then $\mathcal{F}_{\gamma_m,\dots,\gamma_1}$ is the K-vector space whose basis is given by all multipartitions

$$\bigsqcup_{n\geq 0} \left\{\lambda = (\lambda^{(m)}, \dots, \lambda^{(1)}) \vdash n\right\}$$

together with the operators e_i, f_i, t_i, for $0 \le i \le r-1$, and v^d defined by

$$e_i\lambda = \sum_{\mathrm{res}(\lambda/\mu)\equiv i} v^{-N_i^a(\lambda/\mu)}\mu, \quad f_i\lambda = \sum_{\mathrm{res}(\mu/\lambda)\equiv i} v^{N_i^b(\mu/\lambda)}\mu,$$

$$t_i\lambda = v^{N_i(\lambda)}\lambda, \quad v^d\lambda = v^{-W_0(\lambda)}\lambda.$$

THEOREM 10.10. *Let* $\mathrm{res}(x)$ *be the* r*-residue associated with* $(\gamma_m, \dots, \gamma_1)$ *and define* $\mathcal{F}_{\gamma_m,\dots,\gamma_1}$ *as above. Then* $\mathcal{F}_{\gamma_m,\dots,\gamma_1}$ *is an integrable* U_v*-module. The* U_v*-submodule generated by the empty diagram is isomorphic to* $V(\Lambda)$ *where* $\Lambda = \sum_{i=1}^m \Lambda_{\gamma_i}$.

PROOF. Let U'_v be the K-subalgebra of U_v generated by e_i, f_i, t_i $(0 \leq i < r)$. Then we have the following isomorphism of vector spaces by Theorem 6.2(1).

$$U'_v \xrightarrow{\sim} U_v^- \otimes K[t_0^{\pm 1}, \ldots, t_{r-1}^{\pm 1}] \otimes U_v^+$$

Hence the defining relations for U'_v are obtained by dropping the relations which involve v^d. In particular, U'_v has an automorphism

$$\tau : e_i \mapsto e_{i-1},\ f_i \mapsto f_{i-1},\ t_i \mapsto t_{i-1},$$

and we have $U_v = U'_v \otimes K[v^d, v^{-d}]$.

Recall that $\mathcal{F}$ is a U'_v-module by Theorem 10.6. If we twist the action by τ^γ, e_i deletes nodes of r-residue $i - \gamma$ under this new module structure. Hence if we consider a new r-residue associated with γ which is obtained from the old one by adding γ, this new U'_v-module is precisely $\mathcal{F}_\gamma$, and e_i deletes nodes of r-residue i. Now we make $\mathcal{F}_{\gamma_m} \otimes \cdots \otimes \mathcal{F}_{\gamma_1}$ into a U'_v-module by using the coproduct $\Delta(e_i) = 1 \otimes e_i + e_i \otimes t_i^{-1}$, $\Delta(f_i) = f_i \otimes 1 + t_i \otimes f_i$, $\Delta(t_i) = t_i \otimes t_i$. The result is the module $\mathcal{F}_{\gamma_m, \ldots, \gamma_1}$ restricted to U'_v. Hence we have proved that this is a U'_v-module. To check that the U'_v-module structure on $\mathcal{F}_{\gamma_m, \ldots, \gamma_1}$ extends to a U_v-module structure via $v^d \lambda = v^{-W_0(\lambda)} \lambda$, it is enough to verify that $v^d e_i v^{-d} = v^{\delta_{i0}} e_i$, $v^d f_i v^{-d} = v^{-\delta_{i0}} f_i$ and $v^d t_i = t_i v^d$. The proof of this is the same as part (3) of the proof of Theorem 10.6. The proof that $U_v \emptyset$ is an irreducible integrable U_v-module with highest weight Λ is also the same as in Theorem 10.6. $\square$

CHAPTER 11

Description of the crystal graph of $V(\Lambda)$

11.1. A theorem for proving the Misra-Miwa theorem

The purpose of this chapter is to describe the crystal graph of the irreducible integrable highest weight modules of the quantum algebra of type $A^{(1)}_{r-1}$. Following Mathas, we use the terminology "good nodes"in the argument below, which has appeared in Kleshchev's famous modular branching rule for the symmetric group [**sM-K1**]-[**sM-K4**], but the argument itself is due to K.C.Misra and T.Miwa [**mp-MM**].

Another remark is that [**mp-MM**] treats $L(\Lambda_0)$ only, and its higher level analogue [**mp-JMMO**] uses the same underlying space with a different module structure. Our choice of the module structure, as explained in the previous chapter, appears in the study of the modular representation theory of the Hecke algebra of type $G(m,1,n)$. This application is summarized in the next chapter.

We start with the definition of a good node. As before, for a node $x \in \lambda$, $N_i^b(x)$ is the number of addable i-nodes lying strictly below x minus the number of removable i-nodes lying strictly below x.

Definition 11.1. We fix $(\gamma_m,\dots,\gamma_1)$ and let $\mathrm{res}(x)$ be the associated r-residue. A removable i-node x of a (multi-)Young diagram $\lambda = (\lambda^{(m)},\dots,\lambda^{(1)})$ is **i-normal** if the following two conditions are fulfilled.

(1) $N_i^b(x) \le 0$,
(2) If a removable i-node y of λ sits below x, then $N_i^b(y) > N_i^b(x)$.

Let x be an i-normal node. If there is no i-normal node above x then we call x a **i-good** node.

As in the previous chapter, we denote the set of addable i-nodes of λ and the set of removable i-nodes of λ by $A_i(\lambda)$ and $R_i(\lambda)$ respectively.

Lemma 11.2. *Let $\lambda = (\lambda^{(m)},\dots,\lambda^{(1)})$ be a (multi-)Young diagram. Reading the nodes of $A_i(\lambda) \sqcup R_i(\lambda)$ from top to bottom (see Definition 10.7), write down a sequence of A_i's and R_i's corresponding to the addable and removable i-nodes respectively. Delete a pair of adjacent nodes of the form R_iA_i and keep on doing until no such pair remains. Then the i-normal nodes of λ correspond to the R_i's in the resulting sequence. Consequently, the resulting sequence does not depend on the order in which the R_iA_i pairs are deleted.*

Before proving this, we give an example.

Example 11.3. Take $r = 3, (\gamma_2,\gamma_1) = (0,2) \in (\mathbb{Z}/3\mathbb{Z})^2$ and consider the example $\lambda = (541, 3211)$.

$\lambda^{(2)} =$

0	1	2	0	1	2
2	0	1	2		
1	2				

$\lambda^{(1)} =$

2	0	1	2
1	2		
0			
2			

Take $i = 2$. Reading the addable and removable 2-nodes starting from $\lambda^{(1)}$ we get $A_2R_2R_2A_2R_2A_2$. After deleting two R_2A_2 pairs the process ends. The remaining R_2 corresponds to the removable 2-node in the second row of $\lambda^{(1)}$. Call it x. Then $N_2^b(x) = 0$ and the other removable 2-nodes, namely the node in the fourth row of $\lambda^{(1)}$ and the node in the second row of $\lambda^{(2)}$, both have $N_2^b(y) = 1$, which is certainly greater than $N_2^b(x)$. Hence x is 2-normal.

If we consider the (multi-)Young diagram which is obtained by moving the removable node in the second row of $\lambda^{(2)}$ to the first row of $\lambda^{(2)}$ then we get the sequence $A_2R_2R_2R_2A_2A_2$. So the same node x as before is 2-normal.

Now we prove the lemma.

PROOF. Assume that after a finite number of steps we are still able to delete a pair R_iA_i. In this case, we must show that the node correponding to this R_i cannot be i-normal.

Let x and x' be the removable and addable i-nodes corresponding to the R_i and A_i for the pair which we are about to delete.

First we consider the case where there is no removable i-node below x'. Then we have $N_i^b(x) = N_i^b(x') + 1 \geq 1$ and x cannot be i-normal.

Next we consider the case where there exists a removable i-node below x'. Let y be the highest removable i-node below x'. Since we have deleted pairs of a removable i-node and an addable i-node before x and x' become adjacent, the number of addable i-nodes between x and x' is the same as the number of removable i-nodes between x and x'. Further, only addable i-nodes appear between x' and y. Consequently, the number of addable i-nodes in the interval $(x, y]$ in the original sequence is greater than or equal to the number of removable i-nodes in this interval. Therefore, $N_i^b(y) \leq N_i^b(x)$ and x is not i-normal. We have proved that none of the nodes deleted in this process are i-normal.

Next assume that we have reached the stage where no R_iA_i is removable. We must show that the remaining R_i's correspond to i-normal nodes. Note that the final sequence has the form $A_i \cdots A_iR_i \cdots R_i$. Let t be the number of R_i in this sequence. We read the R_i's from left to right and label the corresponding removable i-nodes by $x_1, \cdots, x_t$. Then we have

$$N_i^b(x_1) = -t + 1, \; N_i^b(x_2) = -t + 2, \ldots, N_i^b(x_t) = 0.$$

Hence we have $N_i^b(x) \leq 0$ for $x = x_1, \ldots, x_t$.

If we can prove that $N_i^b(y) > N_i^b(x_j)$ for any removable i-node y which appears between x_j and x_{j+1} then we may conclude that $N_i^b(z) > N_i^b(x_j)$ for any removable i-node z which is below x_j; consequently, $x_1, \ldots, x_t$ are i-normal.

Assume that y is a removable i-node which appears between x_j and x_{j+1}. We consider the sequence of removable i-nodes and addable i-nodes which have been deleted before x_j and x_{j+1} become adjacent. Then y appears in this sequence.

For any N, we consider the first N elements of this sequence and claim that the number of removable i-nodes in these N elements is greater than or equal to the number of addable i-nodes in these N elements, and that if the Nth element is a removable i-node then equality does not hold.

To see this, we record the process of deleting pairs R_iA_i by assigning an opening parethesis and a closing parethesis to the removable i-node and the addable i-node of each pair. For example, we have

$$R_iR_iA_iR_iA_iA_iR_iA_i \Longleftrightarrow (\,(\,)\,(\,)\,)\,(\,).$$

In this example, the number of removable i nodes minus the number of addable i-nodes for $N = 1, \ldots, 8$ are 1 2 1 2 1 0 1 0. The claim is now obvious since the number of opening paretheses always exceeds the number of closing parentheses.

Now assume that the Nth element is y. We denote by $n_r(y)$ the number of removable i-nodes in the first N elements and by $n_a(y)$ the number of addable i-nodes in the first N elements. Then the claim above implies that

$$N_i^b(y) = N_i^b(x_j) - n_a(y) + n_r(y) \geq N_i^b(x_j) + 1.$$

Hence we have proved that all of the x_j are i-normal. □

Before stating the next theorem we fix notation.

DEFINITION 11.4. Let $\mathrm{res}(x)$ be the r-residue associated with $(\gamma_m, \ldots, \gamma_1)$. Suppose that a (multi-)Young diagram λ is given. As in Lemma 11.2, we delete pairs R_iA_i until we get a sequence of the form $A_i \cdots A_iR_i \cdots R_i$. Let s and t be the number of A_i and the number of R_i in this sequence respectively. We denote the corresponding nodes by

$$y_1, \ldots, y_s, x_1, \ldots, x_t,$$

where $y_1, \ldots, y_s \in A_i(\lambda)$ and $x_1, \ldots, x_t \in R_i(\lambda)$. Then set $l = s + t$ and define $\lambda[k]$ $(0 \leq k \leq l)$ by

$$\lambda[k] = \begin{cases} \lambda \setminus \{x_1, \ldots, x_{t-k}\} & (0 \leq k \leq t-1), \\ \lambda & (k = t), \\ \lambda \cup \{y_{l-k+1}, \ldots, y_s\} & (t+1 \leq k \leq l). \end{cases}$$

We also define N_iR_λ by

$$N_iR_\lambda = R_i(\lambda) \setminus \{y_1, \ldots, y_s, x_1, \ldots, x_t\}.$$

This is the set of removable i-nodes of λ which are not i-normal.

For example, we have the following $\lambda[k]$ for Example 11.3.

$$\lambda[0] = (541, 3111), \quad \lambda = \lambda[1] = (541, 3211), \quad \lambda[2] = (541, 4211).$$

DEFINITION 11.5. For each $x \in N_iR_\lambda$, consider nodes $y \in A_i(\lambda)$ which lie below x such that $N_i^b(y) = N_i^b(x) - 1$. We denote the highest such y by $\tilde{x}$.

DEFINITION 11.6. $\mathcal{J}_k$ is the set of permutations of $A_i \cdots A_iR_i \cdots R_i$ in which A_i repeats $l - k$ times and R_i repeats k times. For each $\sigma \in \mathcal{J}_k$, we write $\sigma = \sigma_1 \cdots \sigma_l$ $(\sigma_j = A_i$ or $R_i)$. Then the length $l(\sigma)$ of σ is the number of (j, j') such that $j < j'$ and $\sigma_j = R_i$, $\sigma_{j'} = A_i$.

DEFINITION 11.7. $\lambda(\sigma, k, S)$ is the (multi-)Young diagram associated to a pair (σ, S) of $\sigma \in \mathcal{J}_k$ and a subset S of $N_i R_\lambda$ defined as follows.

- We start with $\lambda[k]$.
- We move k i-normal nodes of $\lambda[k]$ to the nodes which have values R_i in σ.
- We move all $x \in S$ to $\tilde{x}$.
- The resulting diagram is $\lambda(\sigma, k, S)$.

THEOREM 11.8. *Let λ be a (multi-)Young diagram and define $\lambda[k]$ $(0 \leq k \leq l)$ as above. Then we have the following.*

(1) *If $t < k \leq l$ then the i-normal nodes of $\lambda[k]$ are $\{y_{l-k+1}, \dots, y_s, x_1, \dots, x_l\}$. If $0 \leq k \leq t$ then the i-normal nodes are $\{x_{t-k+1}, \dots, x_t\}$.*

(2) *For $x \in N_i R_\lambda$ $\tilde{x}$ is well-defined. Further, $x \neq x'$ implies that $\tilde{x} \neq \tilde{x}'$ and there is no element of $\{y_1, \dots, y_s, x_1, \dots, x_t\}$ between x and $\tilde{x}$. If $z \in N_i R_\lambda$ then either $\{z, \tilde{z}\}$ are both between x and $\tilde{x}$ or both outside this interval.*

(3) *Let $\mathcal{F}_{\gamma_m, \dots, \gamma_1}$ be as in Theorem 10.10 and set*

$$u_k = \sum_{\sigma \in \mathcal{J}_k} \sum_{S \subset N_i R_\lambda} v^{l(\sigma)} (-v)^{|S|} \lambda(\sigma, k, S).$$

Then

$$\tilde{e}_i u_k = \begin{cases} 0 & (k = 0) \\ u_{k-1} & (0 < k \leq l) \end{cases}, \quad \tilde{f}_i u_k = \begin{cases} u_{k+1} & (0 \leq k < l) \\ 0 & (k = l) \end{cases}.$$

(4) *Let $L = \oplus R\lambda$ where λ runs through all multipartitions. Then L is a crystal lattice. That is, L is a full rank R-lattice of $\mathcal{F}_{\gamma_m, \dots, \gamma_1}$, has a weight space decomposition and L is stable under the Kashiwara operators $\tilde{e}_i$ and $\tilde{f}_i$.*

(5) *If $\tilde{e}_i \lambda \not\equiv 0 \pmod{vL}$ then we may write $\tilde{e}_i \lambda \equiv \mu \pmod{vL}$ such that λ/μ is an i-good node of λ and $\lambda \equiv \tilde{f}_i \mu \pmod{vL}$. Conversely, if λ/μ is an i-good node of λ then $\tilde{e}_i \lambda \equiv \mu \pmod{vL}$.*

(6) *If $\tilde{f}_i \lambda \not\equiv 0 \pmod{vL}$ then we may write $\tilde{f}_i \lambda \equiv \mu \pmod{vL}$ such that μ/λ is an i-good node of μ and $\lambda \equiv \tilde{e}_i \mu \pmod{vL}$.*

(7) *If we further set $B = \{\lambda \pmod{vL}\}$ where λ runs through all multipartitions then (L, B) is a crystal basis of $\mathcal{F}_{\gamma_m, \dots, \gamma_1}$.*

PROOF. (1) If we delete pairs $R_i A_i$ from $\lambda[k]$ in the same order as for λ, then we reach the same set of nodes $y_1, \dots, y_s, x_1, \dots, x_t$. Since the first $l - k$ nodes are addable i-nodes and the last k nodes are removable i-nodes, we cannot remove any more pairs $R_i A_i$. Thus we have the result by Lemma 11.2.
(2) Let $A_i \cdots A_i R_i \cdots R_i$ be the final sequence which is obtained from λ by the $R_i A_i$ deletion procedure. Let p and q be two adjacent nodes in the final sequence. We consider the sequence of addable i-nodes and removable i-nodes which are below p and above q: these nodes were deleted before we reach the final sequence. Let

$$\underbrace{R_i \cdots R_i}_{k_1} \underbrace{A_i \cdots A_i}_{l_1} \cdots \underbrace{R_i \cdots R_i}_{k_N} \underbrace{A_i \cdots A_i}_{l_N}$$

be this sequence. As is explained in the proof of Lemma 11.2, the number of R_i among the last k nodes does not exceed the number of A_i among the last k nodes for all k. Hence

$$k_N \leq l_N, \ k_N + k_{N-1} \leq l_N + l_{N-1}, \ \cdots, \ k_N + \cdots + k_1 \leq l_N + \cdots + l_1.$$

(In fact, the last inequality is an equality.)

Following the definition of $\tilde{x}$, we assign $\tilde{x}$ to each $x \in N_i R_\lambda$. If x is one of the last k_N removable i-nodes of the sequence above, then the assignment $x \to \tilde{x}$ is given by the following rule.

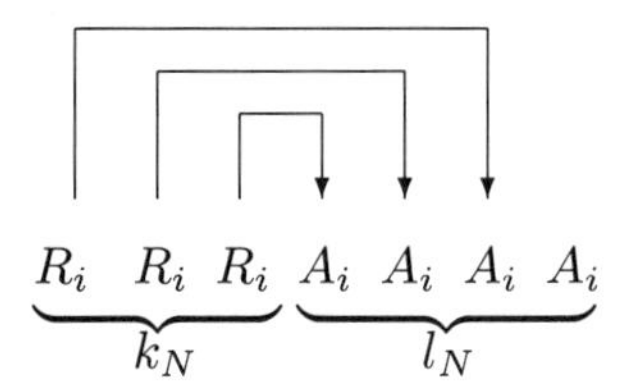

We consider the assignement $x \to \tilde{x}$ for the next k_{N-1} removable i-nodes. If $k_{N-1} \le l_{N-1}$, then the rule is the same; the only difference is that we use k_{N-1}, l_{N-1} instead of k_N, l_N. If $k_{N-1} > l_{N-1}$ then the assignment is done in two steps: first we make the assignment for the last l_{N-1} nodes by the same rule as above. To make the assignment for the remaining $k_{N-1} - l_{N-1}$ nodes, we use nodes among the l_N addable i-nodes which were not used in the assignment for k_N nodes. The condition $k_N + k_{N-1} \le l_N + l_{N-1}$ guarantees that this is possible. This is best understood by an example.

$$\underbrace{R_i\ R_i\ R_i}_{k_{N-1}}\ \underbrace{A_i\ A_i}_{l_{N-1}}\ \underbrace{R_i}_{k_N}\ \underbrace{A_i\ A_i}_{l_N}$$

We continue the assignment in the similar way. If we understand the assignment in this manner, it is clear that we are making pairs of opening and closing paretheses in the assignment procedure. Thus, the assignment $x \to \tilde{x}$ is well-defined for all $x \in N_i R_\lambda$ and the assertions are all obvious.

(3) We shall compute $e_i u_k = \sum_\sigma \sum_S v^{l(\sigma)}(-v)^{|S|} e_i \lambda(\sigma, k, S)$ and prove that $e_i u_k = [l-k+1] u_{k-1}$. Thus this action on u_k coincides with the action of e on V_l defined in Proposition 4.5.

Since $\lambda(\sigma, k, S)$ is obtained from λ first by deleting all nodes in $R_i(\lambda)$ and then adding some of the nodes in

$$\{y_1, \ldots, y_s, x_1, \ldots, x_t\} \sqcup \left(\bigsqcup_{x \in N_i R_\lambda} \{x, \tilde{x}\} \right), \tag{11.1}$$

$R_i(\lambda(\sigma, k, S))$ is contained in the set of (11.1). Similarly, since $\lambda(\sigma, k, S)$ is obtained from λ first by adding all of the nodes in $A_i(\lambda)$ and then deleting some of the nodes in (11.1), $A_i(\lambda(\sigma, k, S))$ is also contained in this set.

Fix $\sigma \in \mathcal{J}_k$ and consider all $\lambda(\sigma, k, S)$ which have this common σ. Since k i-normal nodes of $\lambda[k]$ are moved to the same place in these $\lambda(\sigma, k, S)$, we denote the set of these nodes by $J_{\sigma,k}$. We also denote $\{y_1, \ldots, y_s, x_1, \ldots, x_t\} \setminus J_{\sigma,k}$ by $\overline{J}_{\sigma,k}$.

Then we have

$$\begin{cases} R_i(\lambda(\sigma,k,S)) = J_{\sigma,k} \sqcup \{z|z \notin S\} \sqcup \{\tilde{z}|z \in S\}, \\ A_i(\lambda(\sigma,k,S)) = \overline{J}_{\sigma,k} \sqcup \{z|z \in S\} \sqcup \{\tilde{z}|z \notin S\}. \end{cases}$$

For each $x \in N_iR_\lambda$, we denote by $Y_{\sigma,k}(x)$ the set of (multi-)Young diagrams $\lambda(\sigma,k,S)\setminus\{x,\tilde{x}\}$. If we partition $\{\lambda(\sigma,k,S)\}_{S\subset N_iR_\lambda}$ into two according to whether $x \in S$ or $x \notin S$, then we may write $\{\lambda(\sigma,k,S)\}_{S\subset N_iR_\lambda}$ as a disjoint union

$$\{\mu \cup \{x\} \mid \mu \in Y_{\sigma,k}(x)\} \sqcup \{\mu \cup \{\tilde{x}\} \mid \mu \in Y_{\sigma,k}(x)\}.$$

Therefore, if we set $S_\mu = \{x \in N_iR_\lambda \mid \tilde{x} \in \mu\}$ then $S = S_\mu$ if $\lambda(\sigma,k,S) = \mu \cup \{x\}$ and $S = S_\mu \cup \{x\}$ if $\lambda(\sigma,k,S) = \mu \cup \{\tilde{x}\}$.

Now we consider a node deleted by e_i: it belongs to either a pair $\{x,\tilde{x}\}$ or $J_{\sigma,k}$. Hence we may express $\sum_S(-v)^{|S|}e_i\lambda(\sigma,k,S)$ as follows.

$$\sum_{x\in N_iR_\lambda}\sum_{\mu\in Y_{\sigma,k}(x)}\left((-v)^{|S_\mu|+1}v^{-N_i^a(\tilde{x})} + (-v)^{|S_\mu|}v^{-N_i^a(x)}\right)\mu \tag{11.2}$$
$$+\sum_S(-v)^{|S|}\sum_{\mu:\lambda(\sigma,k,S)/\mu\,\in J_{\sigma,k}} v^{-N_i^a(\lambda(\sigma,k,S)/\mu)}\mu.$$

Note that we have proved in (2) that the nodes which lie between x and $\tilde{x}$ form a collection of pairs of the form $\{z,\tilde{z}\}$. As a consequence, $N_i^a(x)$ takes the same value for the cases $\lambda(\sigma,k,S) = \mu\cup\{x\}$ and $\lambda(\sigma,k,S) = \mu\cup\{\tilde{x}\}$, and $N_i^a(\tilde{x}) = N_i^a(x)+1$ if $\lambda(\sigma,k,S) = \mu\cup\{\tilde{x}\}$. Thus all of the terms in the first sum of (11.2) vanish, and it is enough to consider those $\lambda(\sigma,k,S)$ which satisfy $\lambda(\sigma,k,S)/\mu \in J_{\sigma,k}$. We denote $\lambda(\sigma,k,\emptyset)$ by $\lambda(\sigma,k)$, and define $\mu(\sigma,k)$ by

$$\lambda(\sigma,k,S)/\mu = \lambda(\sigma,k)/\mu(\sigma,k).$$

Then μ is obtained from $\mu(\sigma,k)$ by moving nodes $z \in S$ to $\tilde{z}$. If we denote this μ by $\mu(\sigma,k,S)$, then the correspondence between $\mu(\sigma,k)$ and $\mu(\sigma,k,S)$ is bijective. Since pairs $\{z,\tilde{z}\}$ are both above an i-normal node or both below an i-normal node by (2), we have

$$N_i^a(\lambda(\sigma,k,S)/\mu(\sigma,k,S)) = N_i^a(\lambda(\sigma,k)/\mu(\sigma,k)).$$

Using this we obtain

$$e_iu_k = \sum_{S\subset N_iR_\lambda}(-v)^{|S|}\sum_{\sigma\in\mathcal{J}_k}v^{l(\sigma)}\sum_{\mu(\sigma,k)}v^{-N_i^a(\lambda(\sigma,k)/\mu(\sigma,k))}\mu(\sigma,k,S),$$

where $\mu(\sigma,k)$ are such that $\lambda(\sigma,k)/\mu(\sigma,k) \in J_{\sigma,k}$.

Therefore, the proof that $e_iu_k = [l-k+1]u_{k-1}$ is reduced to showing that

$$\sum_{\sigma\in\mathcal{J}_k}\sum_{\mu(\sigma,k)}v^{l(\sigma)-N_i^a(\lambda(\sigma,k)/\mu(\sigma,k))}\mu(\sigma,k) = [l-k+1]\sum_{\tau\in\mathcal{J}_{k-1}}v^{l(\tau)}\lambda(\tau,k-1). \tag{11.3}$$

To prove this, we assume that the removable i-nodes of $\lambda(\tau,k-1)$ in $\{y_1,\dots,x_t\}$ appear as the $i_1,\dots,i_{k-1}$th nodes of $y_1,\dots,x_t$. For each $\lambda(\tau,k-1)$, we consider those σ which satisfy $\mu(\sigma,k) = \lambda(\tau,k-1)$ for some $\mu(\sigma,k)$. Note that once σ is fixed then $\mu(\sigma,k)$ is uniquely determined. Let us write $\{1,\dots,l\}\setminus\{i_1,\dots,i_{k-1}\} = \{j_1,\dots,j_{l-k+1}\}$. Then these σ are obtained by adding the j_sth node to the $i_1, \ldots, i_{k-1}$th nodes of $y_1, \ldots, x_t$. If we denote this node by j_s for short, and denote the corresponding σ by σ_{j_s}, then

$$l(\sigma_{j_s}) = l(\tau) - |\{i_t \mid i_t < j_s\}| + |\{j_t \mid j_t > j_s\}|,$$
$$N_i^a(j_s) = |\{j_t \mid j_t < j_s\}| - |\{i_t \mid i_t < j_s\}|.$$

As a result, we obtain $l(\sigma_{j_s}) - N_i^a(j_s) = l(\tau)+l-k+2-2s$ and the coefficient of $\lambda(\tau, k-1)$ on the left hand side of (11.3) is

$$v^{l(\tau)} \sum_{s=1}^{l-k+1} v^{l-k+2(1-s)} = v^{l(\tau)}[l-k+1].$$

Next we shall compute $f_i u_k = \sum_\sigma \sum_S v^{l(\sigma)}(-v)^{|S|} f_i \lambda(\sigma, k, S)$ and prove that $f_i u_k = [k+1]u_{k+1}$. This will show that the action of f_i on u_k $(0 \le k \le l)$ coincides with the action of f on V_l defined in Proposition 4.5.

We fix $\sigma \in \mathcal{J}_k$, and denote by $Z_{\sigma,k}(x)$ the set of (multi-)Young diagrams $\lambda(\sigma, k, S) \cup \{x, \tilde{x}\}$. If we set $S_\mu = \{z \in N_i R_\lambda \mid \tilde{z} \in \mu\}$, for $\mu \in Z_{\sigma,k}(x)$, then we have $S = S_\mu$ if $\lambda(\sigma, k, S) = \mu \setminus \{x\}$ and $S = S_\mu \setminus \{x\}$ if $\lambda(\sigma, k, S) = \mu \setminus \{\tilde{x}\}$.

Now we consider a node added by f_i. Then it belongs to either a pair of the form $\{x, \tilde{x}\}$ or $J_{\sigma,k}$. We may express $\sum_S (-v)^{|S|} f_i \lambda(\sigma, k, S)$ as follows.

$$\sum_{x \in NiR_\lambda} \sum_{\mu \in Z_{\sigma,k}(x)} \left((-v)^{|S_\mu|-1} v^{N_i^b(\tilde{x})} + (-v)^{|S_\mu|} v^{N_i^b(x)}\right) \mu$$
$$+ \sum_S (-v)^{|S|} \sum_{\mu:\mu/\lambda(\sigma,k,S) \in J_{\sigma,k}} v^{N_i^b(\mu/\lambda(\sigma,k,S))} \mu.$$

$N_i^b(\tilde{x})$ takes the same value for the cases $\lambda(\sigma, k, S) = \mu \setminus \{x\}$ and $\lambda(\sigma, k, S) = \mu \setminus \{\tilde{x}\}$, and $N_i^b(x) = N_i^b(\tilde{x}) - 1$ if $\lambda(\sigma, k, S) = \mu \setminus \{x\}$. Thus we have cancellation, and the same argument as before leads to the equation

$$f_i u_k = \sum_{S \subset N_i R_\lambda} (-v)^{|S|} \sum_{\sigma \in \mathcal{J}_k} v^{l(\sigma)} \sum_{\mu(\sigma,k)} v^{N_i^b(\mu(\sigma,k)/\lambda(\sigma,k))} \mu(\sigma, k, S).$$

Hence the proof that $f_i u_k = [k+1]u_{k+1}$ is reduced to showing that

$$\sum_{\sigma \in \mathcal{J}_k} \sum_{\mu(\sigma,k)} v^{l(\sigma)+N_i^b(\mu(\sigma,k)/\lambda(\sigma,k))} \mu(\sigma, k) = [k+1] \sum_{\tau \in \mathcal{J}_{k+1}} v^{l(\tau)} \lambda(\tau, k+1). \tag{11.4}$$

To prove this, assume that the removable i-nodes of $\lambda(\tau, k+1)$ in $\{y_1, \ldots, x_t\}$ appear as the $i_1, \ldots, i_{k+1}$th nodes of $y_1, \ldots, x_t$. For each $\lambda(\tau, k+1)$, we consider those σ which satisfy $\mu(\sigma, k) = \lambda(\tau, k+1)$ for some $\mu(\sigma, k)$. Then σ is obtained by deleting the i_sth node from the $i_1, \ldots, i_{k+1}$th nodes of $y_1, \ldots, x_t$. We denote this σ by σ_{i_s}, and set $\{1, \ldots, l\} \setminus \{i_1, \ldots, i_{k+1}\} = \{j_1, \ldots, j_{l-k-1}\}$. Then

$$l(\sigma_{i_s}) = l(\tau) + |\{i_t \mid i_t < i_s\}| - |\{j_t \mid j_t > i_s\}|,$$
$$N_i^b(i_s) = |\{j_t \mid j_t > i_s\}| - |\{i_t \mid i_t > i_s\}|.$$

Thus we have $l(\sigma_{i_s}) + N_i^b(i_s) = l(\tau)-k+2s-2$, and the coefficient of $\lambda(\tau, k+1)$ on the left hand side of (11.4) is

$$v^{l(\tau)} \sum_{s=1}^{k+1} v^{-k+2(s-1)} = v^{l(\tau)}[k+1].$$

We have proved $e_i u_k = [l-k+1]u_{k-1}$ and $f_i u_k = [k+1]u_{k+1}$. As a consequence, we have

$$\tilde{e}_i u_k = \begin{cases} 0 & (k=0) \\ u_{k-1} & (1 \le k \le l) \end{cases}, \quad \tilde{f}_i u_k = \begin{cases} u_{k+1} & (0 \le k \le l-1) \\ 0 & (k=l) \end{cases}.$$

(4) We compute u_k for all k for all λ, and denote by L' the R-submodule of L generated by these u_k. As $u_k \equiv \lambda[k] \pmod{vL}$, we have $L = L' + vL$. Thus Nakayama's Lemma implies that $L = L'$. Since we proved in (3) that L' is stable under the Kashiwara operators $\tilde{e}_i$ and $\tilde{f}_i$, the result follows.
(5) By (4), the result in (3) implies that, modulo vL,

$$\tilde{e}_i \lambda[k] \equiv \begin{cases} \lambda[k-1] & (k>0) \\ 0 & (k=0) \end{cases}, \quad \tilde{f}_i \lambda[k] \equiv \begin{cases} \lambda[k+1] & (k<l) \\ 0 & (k=l) \end{cases}.$$

Since $\lambda[k-1]$ is obtained from $\lambda[k]$ by deleting an i-good node, the first part is clear. Since there is at most one i-good node of λ, that λ/μ is a i-good node implies that $\lambda = \lambda[t], \mu = \lambda[t-1]$. Hence the second part is also clear.
(6) The proof is the same as that of (5).
(7) It is obvious that both L and B have weight space decompositions and B is a basis of L/vL. By (3) and (4), L and $B \cup \{0\}$ are stable under the Kashiwara operators $\tilde{e}_i$ and $\tilde{f}_i$. That $\tilde{e}_i\lambda \equiv \mu \Leftrightarrow \lambda \equiv \tilde{f}_i\mu$ follows from (5) and (6). □

11.2. The Misra-Miwa theorem

Now we are in a position to prove the Misra-Miwa theorem. Similar to the definition of $\lambda \vdash n$ in Definition 10.1, if λ is a (multi-)Young diagram with n nodes then we write $\lambda \vdash n$.

DEFINITION 11.9. Let $\lambda \vdash n$. A **standard tableau** of shape λ is a labelling of the nodes of λ by the integers $1, \dots, n$ such that

- The integers $1, \dots, n$ appear exactly once.
- They increase from left to right in each row of each component.
- They increase from top to bottom in each column of each component.

An increasing sequence of (multi-)Young diagrams

$$\emptyset \subset \lambda_{\le 1} \subset \cdots \subset \lambda_{\le n} = \lambda$$

may be identified with a standard tableau by the rule $\lambda_{\le k}/\lambda_{\le k-1} = \boxed{\text{k}}$.

DEFINITION 11.10. Fix $(\gamma_m, \dots, \gamma_1)$ and its associated r-residue as before. A (multi-)Young diagram λ is **Kleshchev** with respect to $(\gamma_m, \dots, \gamma_1)$ if there exists a standard tableau $\mathbf{t}$ of shape λ whose corresponding sequence of (multi-)Young diagrams $\{\lambda_{\le k}\}_{1 \le k \le n}$ has the property that

for all k, the node $\boxed{\text{k}}$ is a i-good node of $\lambda_{\le k}$ for some i.

We denote the set of Kleshchev multipartitions by $\mathcal{KP}_{\gamma_m,\dots,\gamma_1}$.

In the following theorem, which we call the Misra-Miwa theorem, we describe the crystal graph of an irreducible integrable highest weight U_v-module with the notion of Kleshchev multipartitions.

THEOREM 11.11. *Fix $(\gamma_m, \dots, \gamma_1)$ and its associated r-residue. Let $\mathcal{F}_{\gamma_m,\dots,\gamma_1}$ be as in Theorem 10.10, and identify $U_v\emptyset$ with $V(\Lambda)$, where $\emptyset$ is the empty diagram and $\Lambda = \sum_{i=1}^m \Lambda_{\gamma_i}$. Let (L, B) be the crystal basis of $\mathcal{F}_{\gamma_m,\dots,\gamma_1}$ given by Theorem 11.8(7), and set*

$$L_0 = L \cap V(\Lambda), \quad B_0 = B \cap (L_0/vL_0).$$

Then (L_0, B_0) coincides with the crystal basis $(L(\Lambda), B(\Lambda))$, and we have

$$B_0 = \{\lambda \pmod{vL} \mid \lambda \in \mathcal{KP}_{\gamma_m,\dots,\gamma_1}\}.$$

PROOF. By (step 3) in the proof of Theorem 9.6, if $u \in V(\Lambda)$ satisfies $\tilde{e}_i u \in L(\Lambda)$ $(0 \le i < r)$ then $u \in L(\Lambda)$. By Theorem 11.8(4), L_0 is stable under the Kashiwara operators $\tilde{e}_i, \tilde{f}_i$ and $L \cap K\emptyset = R\emptyset$. Thus $L(\Lambda) \subset L_0$ and $L(\Lambda) = L_0$ by downward induction on the weights of $V(\Lambda)$. Next we embed $B(\Lambda) \subset L_0/vL_0$ into L/vL by sending $b \pmod{vL_0}$ to $b \pmod{vL}$. We shall prove that

$$\{b \pmod{vL} | b \in B(\Lambda)\} = \{\lambda \pmod{vL} | \lambda \in \mathcal{KP}_{\gamma_m,\dots,\gamma_1}\} \tag{11.5}$$

by downward induction on the weights of $V(\Lambda)$.

Assume that $\tilde{f}_i \tilde{f}_{i_2} \cdots \tilde{f}_{i_N} \emptyset \pmod{vL(\Lambda)} \in B(\Lambda)$. Then the induction hypothesis implies that there exists $\mu \in \mathcal{KP}_{\gamma_m,\dots,\gamma_1}$ such that

$$\tilde{f}_{i_2} \cdots \tilde{f}_{i_N} \emptyset \equiv \mu \pmod{vL}.$$

If $\tilde{f}_i \mu \equiv 0 \pmod{vL}$ then we have

$$\tilde{f}_i \tilde{f}_{i_2} \cdots \tilde{f}_{i_N} \emptyset \in (vL) \cap V(\Lambda) = vL_0 = vL(\Lambda),$$

which contradicts the assumption that $\tilde{f}_i \tilde{f}_{i_2} \cdots \tilde{f}_{i_N} \emptyset \pmod{vL(\Lambda)}$ belongs to $B(\Lambda)$. Hence we have $\tilde{f}_i \mu \not\equiv 0 \pmod{vL}$. We apply Theorem 11.8(6) to conclude that

$$\tilde{f}_i \tilde{f}_{i_2} \cdots \tilde{f}_{i_N} \emptyset \equiv \lambda \pmod{vL}$$

for some $\lambda \in \mathcal{KP}_{\gamma_m,\dots,\gamma_1}$.

To prove the converse, we assume that $\lambda \in \mathcal{KP}_{\gamma_m,\dots,\gamma_1}$. Then there exists an index i and $\mu \in \mathcal{KP}_{\gamma_m,\dots,\gamma_1}$ such that λ/μ is an i-good node of λ. Now Theorem 11.8(5) implies that $\tilde{e}_i \lambda \equiv \mu \pmod{vL}$, and the induction hypothesis implies that there exists $b \in B(\Lambda)$ such that we may write $\mu \equiv b \pmod{vL}$. Therefore we have

$$\lambda \equiv \tilde{f}_i \mu \equiv \tilde{f}_i b \pmod{vL}.$$

Define $b' = \tilde{f}_i b$. Then $b' \neq 0$, which implies that $b' \in B(\Lambda)$. Thus $\lambda \pmod{vL} = b' \pmod{vL}$ for $b' \in B(\Lambda)$.

Since $B(\Lambda) \pmod{vL}$, the right hand side of (11.5), is a subset of B_0 and a basis of L_0/vL_0, it follows that $B(\Lambda) \pmod{vL}$ and the left hand side of (11.5) must coincide with B_0. $\square$

CHAPTER 12

An overview of the applications to Hecke algebras

12.1. The Hecke algebra of type $G(m,1,n)$

In the remaining chapters I shall explain applications of the theory developed so far to the representation theory of the Hecke algebra of type $G(m,1,n)$: see also [**cr-A**]. We start with the definition of the algebra.

DEFINITION 12.1. Let $\mathbb{S}$ be a commutative associative ring with unit. Given $v_1,\dots,v_m$, and an invertible q in $\mathbb{S}$, the **Hecke algebra** $\mathcal{H}_{n,\mathbb{S}}$ of type $G(m,1,n)$ with the parameters $(q,v_1,\dots,v_m)$ is the $\mathbb{S}$-algebra defined by the following generators and relations.

Generators: a_i $(1\le i\le n)$.
Relations:

$$(a_1-v_1)\cdots(a_1-v_m)=0,\quad (a_i-q)(a_i+1)=0\quad (i\ge 2),$$
$$a_1a_2a_1a_2=a_2a_1a_2a_1,\quad a_ia_j=a_ja_i\ (j\ge i+2),$$
$$a_ia_{i-1}a_i=a_{i-1}a_ia_{i-1}\ (3\le i\le n).$$

We call this the Hecke algebra over $\mathbb{S}$, and n is called its rank.

This algebra was introduced by Malle and the author and its theory has been developed by us and our collaborators since then. See [**H-AK**] and [**H-BM**].

My motivation for introducing this algebra was to generalize the Hecke algebra of type B. On the other hand, Broué's group in Paris had evidence that the Hecke algebra of a complex reflection group should exist through their study of the modular representation theory of finite classical groups of Lie type; see [**cr-BMM**] and [**cH-BMR**]. Guided by this, they introduced the same algebra. My first result about the representation theory of this algebra is a necessary and sufficient condition for semisimplicity and an explicit combinatorial description of the simple modules in this case. These results are explained in the next chapter.

I remark that Cherednik also introduced this algebra [**H-C**]. But he lacked a combinatorial framework within which to construct its simple modules, nor did he think of studying its modular representations. However, it is interesting that this algebra has already appeared in a different context.

Since I found a satisfactory description of the ordinary representations of this algebra, I started the study of its modular representation thory. Luckily, I heard a talk by Leclerc about his work with Lascoux and Thibon. He explained that they had found an algorithm which they conjectured computed the decomposition numbers of the Hecke algebra of type A at roots of unity. This conjecture is now called the LLT conjecture and it was proved by the author. The LLT algorithm actually computes Kashiwara's global basis of $V(\Lambda_0)$; consequently it suggested a link between the decomposition numbers of Hecke algebras and the canonical basis.

The LLT algorithm uses the notion of ladders, which date back to at least James and Kerber [**JK**, 6.3].

DEFINITION 12.2. A partition μ is r-**restricted** if $\mu_{i-1} - \mu_i < r$ for all i. Let μ be an r-restricted partition. Then a **ladder** is the set of nodes (x, y) of μ which lie on a line of the form $y = (1-r)x + r + k$ for some k.

For example, if $\mu = (4, 2, 1)$ and $r = 3$, then the sets of nodes

$$\{(1,1)\}, \{(1,2)\}, \{(2,1),(1,3)\}, \{(2,2),(1,4)\}, \{(3,1)\}$$

are the ladders of μ for $k = 0, 1, 2, 3, 4$ respectively.

Each ladder has a unique residue $i_k := k \pmod r$. We denote the number of nodes in the ladder by n_k and call n_k the size of the ladder.

Let μ be an r-restricted partition as above, and i_k and n_k $(k = 0, \ldots, N)$ the residues and sizes of the ladders. Define an element $A(\mu)$ in the Hayashi realization $\mathcal{F}$ (see Theorem 10.6) by

$$A(\mu) = f_{i_N}^{(n_N)} \cdots f_{i_1}^{(n_1)} \emptyset.$$

As is proved in [**cH-LLT**, Lemma 6.4], [**JK**, 6.3.54] implies that

- μ appears in $A(\mu)$ with coefficient 1.
- All of the partitions which appear in $A(\mu)$ are greater than or equal to μ in the dominance ordering.

If we identify $V(\Lambda_0)$ with the $\mathfrak{g}(A_{r-1}^{(1)})$-submodule of $\mathcal{F}$ generated by the empty partition, then $\{A(\mu) | \mu{:}r\text{-restricted}\}$ is a basis of $V(\Lambda_0)$. Now we compute the canonical basis $G(\mu)$ for r-restricted partitions (=Kleshchev partitions) inductively. We start with the greatest r-restricted partition $\mu_{\max}$ in the dominance ordering and set $G(\mu_{\max}) = A(\mu_{\max})$. Assume that we have determined all $G(\nu)$ for $\nu \rhd \mu$ and that we may write

$$A(\mu) = \sum_{\lambda \trianglerighteq \mu} a_{\lambda,\mu}(v)\lambda, \quad G(\nu) = \sum_{\lambda \trianglerighteq \nu} g_{\lambda,\nu}(v)\lambda,$$

$$G(\mu) = A(\mu) - \sum_{\nu \rhd \mu} b_{\nu,\mu}(v) G(\nu).$$

Then we have

$$g_{\lambda,\mu}(v) = a_{\lambda,\mu}(v) - \sum_{\mu \lhd \nu \trianglelefteq \lambda} g_{\lambda,\nu}(v) b_{\nu,\mu}(v).$$

We require the $g_{\lambda,\mu}(v)$ to be polynomials and $g_{\mu,\mu}(v) = 1$. Thus we must have

$$(a_{\lambda,\mu}(v))_- = \sum_{\mu \lhd \nu \lhd \lambda} (g_{\lambda,\nu}(v) b_{\nu,\mu}(v))_- + (b_{\lambda,\mu}(v))_- \tag{12.1}$$

for r-restricted partitions λ, μ, ν where $p(v) \mapsto (p(v))_-$ is the projection to the second component in the decomposition

$$\mathbb{Q}[v, v^{-1}] \longrightarrow \mathbb{Q}[v] \oplus v^{-1}\mathbb{Q}[v^{-1}].$$

Equation (12.1) combined with the requirement that $b_{\lambda,\mu}(v) = b_{\lambda,\mu}(v^{-1})$ determines $b_{\lambda,\mu}(v)$ inductively. This is the LLT algorithm for computing $G(\mu)$. The LLT conjecture predicts that if we write $G(\mu) = \sum d_{\lambda,\mu}(v)\lambda$, then the coefficients $d_{\lambda,\mu}(1)$ are decomposition numbers for the Hecke algebra of type A with $q = \sqrt[r]{1} \in \overline{\mathbb{Q}}$ $(r \geq 2)$.

When I listened to Leclerc's talk, I knew that the Hecke algebras of type A are a special case of the Hecke algebras of type $G(m, 1, n)$ and that these algebras are quotients of the affine Hecke algebras of type A. Furthermore, the same perverse sheaves appear in the geometric representation theory of both the affine Hecke algebras and the quantum algebras. Because of this I soon realized why the LLT conjecture should be true and I was able to generalize it to the Hecke algebras of type $G(m, 1, n)$: see Theorem 12.5 below. We start with notation.

DEFINITION 12.3. Let $\mathbb{F}$ be a field. Then $\mathcal{H}_{n,\mathbb{F}} - mod$ is the category of finitely generated left $\mathcal{H}_{n,\mathbb{F}}$-modules.

Consider the sequence of the Hecke algebras $\mathcal{H}_{n,\mathbb{F}}$ for $n = 1, 2, \ldots$ with common parameters. Then $\mathcal{H}_{n,\mathbb{F}}$ is a subalgebra of $\mathcal{H}_{n+1,\mathbb{F}}$ and

$$\begin{aligned} \mathrm{Ind} &: \mathcal{H}_{n,\mathbb{F}} - mod \to \mathcal{H}_{n+1,\mathbb{F}} - mod \\ \mathrm{Res} &: \mathcal{H}_{n,\mathbb{F}} - mod \to \mathcal{H}_{n-1,\mathbb{F}} - mod \end{aligned}$$

are the induction and the restriction functors.

Let $\mathcal{H}_{n,\mathbb{F}} - proj$ be the full subcategory of $\mathcal{H}_{n,\mathbb{F}} - mod$ consisting of projective $\mathcal{H}_{n,\mathbb{F}}$-modules. We denote by $V_n(\mathbb{F})$ the complex vector space obtained from the Grothendieck group by tensoring with $\mathbb{C}$. That is,

$$V_n(\mathbb{F}) = K_0(\mathcal{H}_{n,\mathbb{F}} - proj) \otimes_{\mathbb{Z}} \mathbb{C}.$$

We set $V(\mathbb{F}) = \bigoplus_{n \geq 0} V_n(\mathbb{F})$.

As is explained in the next chapter, if we define elements $t_1, \ldots, t_n$ by $t_1 = a_1$ and $t_i = q^{-1} a_i t_{i-1} a_i$ for $i > 1$ then the t_i pairwise commute, and symmetric polynomials in these elements are central in $\mathcal{H}_{n,\mathbb{F}}$. In particular, if we set $c_n(z) = \prod_{i=1}^{n} (z - t_i)$, the coefficients are central.

DEFINITION 12.4. Let M be a $\mathcal{H}_{n,\mathbb{F}}$-module and suppose that $\lambda(z)$ is a monic polynomial of degree n with coefficients in $\mathbb{F}$. Then $P_{n,\lambda(z)}(M)$ is the generalized eigenspace of eigenvalue $\lambda(z)$ of $c_n(z)$ acting on M.

THEOREM 12.5 ([**cH-A2**, Theorem 4.4]). *Let $\mathbb{F}$ be a field and consider the sequence of the Hecke algebras $\mathcal{H}_{n,\mathbb{F}}$ for $n = 1, 2, \ldots$ with common parameters*

$$q = \sqrt[r]{1} \neq 1, \quad v_i = q^{\gamma_i} \ (\gamma_i \in \mathbb{Z}/r\mathbb{Z}, 1 \leq i \leq m). \tag{12.2}$$

Let n_i be the multiplicity of q^i in $\{v_1, \ldots, v_m\}$ and set $\Lambda = \sum_{i=0}^{r-1} n_i \Lambda_i$, where Λ_i are the fundamental weights of the Kac-Moody Lie algebra of type $A^{(1)}_{r-1}$. Then $V(\mathbb{F})$ affords the module structure of the Kac-Moody Lie algebra of type $A^{(1)}_{r-1}$ where the action of the Chevalley generators e_i and f_i is given by

$$e_i[M] = [\bigoplus_{\lambda(z) \in \mathbb{F}[z]} P_{n-1,\lambda(z)/(z-q^i)} \left(\mathrm{Res}(P_{n,\lambda(z)}(M))\right)],$$

$$f_i[M] = [\bigoplus_{\lambda(z) \in \mathbb{F}[z]} P_{n+1,\lambda(z)(z-q^i)} \left(\mathrm{Ind}(P_{n,\lambda(z)}(M))\right)].$$

Moreover, $V(\mathbb{F})$ is isomorphic to the irreducible highest weight module with highest weight Λ, and if the characteristic of $\mathbb{F}$ is 0, then the canonical basis of $V(\mathbb{F})$ specialized at $v = 1$ coincides with the basis of $V(\mathbb{F})$ given by indecomposable projective modules.

The operators e_i and f_i are the i-restriction and the i-induction operators introduced by the author. These operators are natural analogues of the translation functors. Notice that induction followed by taking generalized eigenspaces has been used in representation theory for a long time. The point is that these modified translation functors for the Hecke algebras make $V(\mathbb{F})$ into a module for a Kac-Moody algebra.

This theorem also allows us to introduce a crystal graph structure on the union of the sets of simple $\mathcal{H}_{n,\mathbb{F}}$-modules for $n \geq 0$. The structure does not depend on the characteristic of $\mathbb{F}$ as is proved in [**cH-AM1**]. I also remark that the original LLT algorithm now has variations: the modified LLT algorithm of Goodman and Wenzl [**qu-GW**], and the LT algorithm by Leclerc and Thibon [**qu-LT1**][**qu-LT2**]. The last one is particularly important, because Uglov has generalized the LT algorithm to higher level Fock spaces [**qu-U**]; this gives us an algorithm to compute the decomposition numbers of $\mathcal{H}_{n,\mathbb{F}}$ in characteristic 0 (see also [**cH-A3**]). If the reader is interested in the development of the theory after [**cH-A2**], I recommend reading [**cr-A**].

The detailed proof of Theorem 12.5 begins in the next chapter; here I give the general strategy of the proof. First, we show the surjectivity of the decomposition map and a concrete (i-)restriction rule. This will imply that if $V(\mathbb{F})$ is a cyclic $U(\mathfrak{g}(A^{(1)}_{r-1}))$-module then all the statements except the canonical basis part hold. Thus we are reduced to showing that $V(\mathbb{F})$ is cyclic and that the canonical basis corresponds to the basis consisting of PIMs. Here we use the fact that the Hecke algebra of type $G(m, 1, n)$ is a quotient algebra of the (extended) affine Hecke algebra of type A and apply results from geometric representation theory. Specifically, we use a geometric construction of the standard modules of the affine Hecke algebra and its relation to a geometric construction of the quantum algebra and the Hall algebra.

The standard module has two constructions. One using topological K-theory due to Kazhdan-Lusztig, in which the Kazhdan-Lusztig induction theorem is proved. The other using the K-theory of equivariant coherent sheaves due to Ginzburg, in which the multiplicities of irreducible modules in standard modules are given by perverse sheaves. It is known that these two constructions yield modules with the same composition factors, as is explained in [**cH-A2**, Theorem 3.2]. We only cite these results in Chapter 14 and do not explain the ideas of the proof. On the other hand, we do explain the geometric construction of the quantum algebra and the Hall algebra in some detail, as was promised in Chapter 7.

Together with some technical calculations, these facts allow us to relate $V(\mathbb{F})$ to an integrable U_v-module specialized at $v = 1$. As a result, we reach Theorem 12.5.

We remark that Ginzburg's theorem does not give Theorem 12.5 in a direct way since what we need is the relation between Specht modules and simple modules, not the relation between standard modules and simple modules. To bypass this difficulty, we keep track of the $U^-(\mathfrak{g}(A^{(1)}_{r-1}))$-equivariance in the course of the proof by introducing i-restriction for the affine Hecke algebra; this also defines the $U^-(\mathfrak{g}(A^{(1)}_{r-1}))$ action. This is one of the key ideas in the proof.

As we explained, Theorem 12.5 sprang from the LLT algorithm, which is a very elementary and explicit combinatorial algorithm: yet its verification relies on deep results of Lusztig and Ginzburg, which use intersection cohomology, a sophisticated tool from mainstream mathematics which was developed for proving the Weil conjectures. So, the LLT algorithm started as a kind of applied mathematics (combinatorics) and evolved into something richer. In this sense, this theorem has an interesting history. Even after its birth, a lot of combinatorics and representation theory has played a role in its development.

I refer the reader to [**cb-A**]-[**cb-F**], [**CG**] and [**aH-G1**]-[**aH-L2**] for the references for the geometry needed in the proof of Theorem 12.5; see also Nakajima's papers [**cb-N1**] and [**cb-N2**] for results which are intimately related to Lusztig's work on quantum algebras.

12.2. Consequences of Theorem 12.5

We have proved in the previous chapters that the canonical basis induces the crystal basis. We have also described the crystal graph in terms of Kleshchev multipartitions. Now we explain how these results are applied.

First of all, we can determine a complete set of simple modules for the Hecke algebra of type $G(m,1,n)$ in arbitrary characteristic. Recall that there is a Specht module theory for the cyclotomic Hecke algebra; see Dipper, James and Mathas [**S-DJM2**] and the next chapter. Each Specht module is equipped with an invariant form and if we factor out by the radical of this form we get a module which is either absolutely irreducible or zero.

Let $D^\lambda = S^\lambda/\mathrm{rad}_{\langle\ ,\ \rangle} S^\lambda$ be the module associated with $\lambda = (\lambda^{(m)},\dots,\lambda^{(1)})$. Then the non-zero D^λ form a complete set of irreducible modules. The last statement is proved using [**S-GL**]. In fact, the authors of [**S-GL**] have already applied their result to our Hecke algebra; however, the module they use is a cell module, which is not a natural generalization of the classical Specht module of the symmetric group or the Hecke algebra of type A.

We cannot go further using this Specht module theory alone, but we may determine exactly which D^λ are non-zero by using Theorem 12.5 and a result in [**qu-U**]. See [**cH-A3**] for the details. The theorem is as follows. Recall that $\mathcal{K}P_{\gamma_m,\dots,\gamma_1}$ is the set of Kleshchev multipartitions defined for the r-residue associated with $(\gamma_m,\dots,\gamma_1)$.

THEOREM 12.6 ([**cH-A3**, Theorem 4.3]). *Let $\mathbb{F}$ be a field and assume the parameter condition (12.2) of Theorem 12.5. Then a complete set of pairwise non-isomorphic irreducible modules of the Hecke algebra of type $G(m,1,n)$ is given by*

$$\{D^\lambda | \lambda \in \mathcal{K}P_{\gamma_m,\dots,\gamma_1}\}.$$

In this way, the description of the crystal graph developed in the last chapter is applied. The parameter condition (12.2) is not a restriction by virtue of a Morita equivalence theorem of Dipper and Mathas, as is explained in the next chapter. The reader may think that this theorem does not give the number of irreducible modules; however, it is also easy to obtain the generating function for the number of Kleshchev multipartitions. This is a consequence of the crystal graph description in Chapter 11; see [**cH-AM1**]. Theorem 12.6 was a new result even for the Hecke algebra of type B: no formula was known for the number of irreducible modules.

Surprisingly, the rule for the Kashiwara operators acting on these crystal graphs is the same as Kleshchev's modular branching rule; this is the reason why we use the terminology "Kleshchev multipartition". The link between the modular branching rule and the crystal graph was first observed by Kleshchev. We note here that an abstract crystal was given recently by Vazirani and Grojnowski in Vazirani's thesis. For an explicit derivation of the crystal properties of this abstract crystal see Brundan and Kleshchev [**sM-BK**]. We also note that when we speak of the modular branching rule it has a specific meaning: the modular branching rule is a statement using good nodes of Young diagrams. For the symmetric group case, see Kleshchev's series of papers [**sM-K1**]-[**sM-K4**]; for the Hecke algebra of type A, see [**sM-B1**].

As other application of Theorem 12.5 we also have the following result.

THEOREM 12.7 ([**cH-AM1**, Theorem B]). *The isomorphism classes of finite dimensional irreducible representations of the affine Hecke algebra of type A over* $\mathbb{F}$ *are indexed by (tuples of) aperiodic multisegments; or equivalently, by (tuples of) restricted Deligne-Langlands parameters.*

We do not explain the terminology here; see [**cH-AM1**] for details. This result confirms a conjecture of Vigneras [**na-V3**] about the classification of the modular irreducible representations of the general linear group over a local field. She had proved almost all parts of her conjecture, but lacked the classification of the simple modules of affine Hecke algebras of type A in positive characteristic. We were lucky to be able to contribute to the last step of Vigneras' program. We remark that Vigneras has her own argument to replace part of this theorem, but she still needs our Hecke algebra to prove this.

In the following, $\langle(s,N)\rangle$ stands for the irreducible (smooth) representation introduced by Vigneras, which generalizes Zelevinski's construction for the characteristic 0 case; see [**na-V4**] for details.

THEOREM 12.8 (Vigneras). *Let F be a local field with residue field of characteristic p, and let k be an algebraically closed field of characteristic $l \neq p$. Let $\hat{G}_{uni}$ be the unipotent admissible dual of $G = GL(n,F)$ over k. Then*

$$\hat{G}_{uni} = \{\langle(s,N)\rangle\},$$

where (s,N) runs through the Deligne-Langlands parameters.

Here the (s,N) are in bijection with unipotent supercuspidal Zelevinski parameters. If we consider cuspidal parameters instead of supercuspidal parameters, they are in bijection with unipotent restricted Zelevinski parameters.

Malle has generalized the theory of symbols, generic degrees, and Green functions to the Hecke algebras of type $G(m,1,n)$. There are also Hecke algebras for the other types of complex reflection groups coming from generalized KZ equations. These are the cyclotomic Hecke algebras. Because of Malle's work (and Broué, Michel, ...), we expect that algebras with good properties which generalize the classical groups of Lie type are hidden behind those results and formulas.

The modular representation theory of the Hecke algebras was originally developed by Dipper and James who were interested in applying it to the modular representation theory of the finite classical groups of Lie type. [**mo-DJHM**] is a good reference to learn about the developments in this direction.

I started the study of Hecke algebras mainly because I was interested in this theory. I like this concrete world in which both combinatorics and representation theory combine to create a wonderful tapestry. I hope that you too feel that this field is charming.

Now we start the proof of Theorem 12.5 mostly following [**cH-A2**]; to make it more accesable to the reader, I have also included the necessary background material. A big change from the original paper is that we adopt Dipper-James-Mathas' Specht module theory in the proof. This is not only because it agrees with our tastes but also because I believe that this theory provides us with the most natural framework.

CHAPTER 13

The Hecke algebra of type $G(m, 1, n)$

13.1. The affine Hecke algebra

DEFINITION 13.1. Let $\mathbb{S}$ be a ring and $q \in \mathbb{S}^\times$ be an invertible element. The **affine Hecke algebra** $\hat{H}_{n,\mathbb{S}}$ (of type A) is the $\mathbb{S}$-algebra defined by the following generators and relations.

Generators: $X_1^{\pm 1}, \ldots, X_n^{\pm 1}, T_2, \ldots, T_n$,
Relations:

$$X_iX_i^{-1} = X_i^{-1}X_i = 1, \ X_iX_j = X_jX_i \ (1 \le i, j \le n),$$
$$T_iT_{i-1}T_i = T_{i-1}T_iT_{i-1} \ (3 \le i \le n),$$
$$T_iT_j = T_jT_i \ (j \ge i+2),$$
$$(T_i - q)(T_i + 1) = 0 \ \ (2 \le i \le n),$$
$$T_iX_{i-1}T_i = qX_i \ (2 \le i \le n), \ \ T_iX_j = X_jT_i \ (j \ne i-1, i).$$

Let $P = \oplus_{i=1}^n \mathbb{Z}\epsilon_i$ be the free $\mathbb{Z}$-module with basis $\{\epsilon_i\}_{1 \le i \le n}$. The symmetric group S_n acts on P by $w\epsilon_i = \epsilon_{w(i)}$ $(w \in S_n)$. We let $s_i = (i-1, i)$ be the transposition which interchanges $i-1$ and i and keeps all other indices fixed. Then, $\{s_i\}_{2 \le i \le n}$ generate S_n and the following relations are the defining relations for S_n.

$$s_i^2 = 1, \ \ s_is_{i-1}s_i = s_{i-1}s_is_{i-1}, \ \ s_is_j = s_js_i \ (j \ge i+2).$$

For each $w \in S_n$, we consider expressions of the form $w = s_{i_1} \cdots s_{i_l}$. A **reduced expression** for w is an expression of w whose proper subwords are expressions of elements other than w; equivalently, l is minimal such that $w = s_{i_1} \cdots s_{i_l}$. Call l the **length** of w and write $l(w) = l$.

Set $T_w = T_{i_1} \cdots T_{i_l}$ when $w =_{i_1} \cdots s_{i_l}$ is a reduced expression. It is known that T_w depends only on w and does not depend on the choice of a reduced expression. The argument to prove that T_w is independent of the choice of a reduced expression relies only on the braid relations, consequently, it also works for the Hecke algebra of type $G(m, 1, n)$. See [**GP**, Lemma 4.4.3] for this argument. Hence we may define $a_w = a_{i_1} \cdots a_{i_{l(w)}} \in \mathcal{H}_{n,\mathbb{S}}$ by choosing a reduced expression $w = s_{i_1} \cdots s_{i_{l(w)}}$.

The following formulas for $i \ge 2$ follows easily from the relations in $\hat{H}_{n,\mathbb{S}}$.

$$T_iT_w = \begin{cases} T_{s_iw} & (l(s_iw) > l(w)) \\ qT_{s_iw} + (q-1)T_w & (l(s_iw) < l(w)) \end{cases}$$

$$T_wT_i = \begin{cases} T_{ws_i} & (l(ws_i) > l(w)) \\ qT_{ws_i} + (q-1)T_w & (l(ws_i) < l(w)) \end{cases}$$

The same argument also gives the following.

$$a_i a_w = \begin{cases} a_{s_i w} & (l(s_i w) > l(w)) \\ q a_{s_i w} + (q-1)a_w & (l(s_i w) < l(w)) \end{cases}$$

$$a_w a_i = \begin{cases} a_{w s_i} & (l(w s_i) > l(w)) \\ q a_{w s_i} + (q-1)a_w & (l(w s_i) < l(w)) \end{cases}$$

Let $\alpha_i = -\epsilon_{i-1} + \epsilon_i$. Given $\lambda = \sum_{i=1}^n \lambda_i \epsilon_i \in P$ define X_λ by $X_\lambda = X_1^{\lambda_1} \cdots X_n^{\lambda_n}$. Then the last two relations in the definition of the affine Hecke algebra may be replaced with

$$T_i X_\lambda = X_{s_i\lambda} T_i + (q-1)\frac{X_\lambda - X_{s_i\lambda}}{1 - X_{-\alpha_i}}.$$

This presentation of $\hat{H}_{n,\mathbb{S}}$ is called the **Bernstein presentation**. The following is well-known.

- $\{X_\lambda T_w\}_{\lambda\in P, w\in S_n}$ is an $\mathbb{S}$-basis of $\hat{H}_{n,\mathbb{S}}$.
- The center $Z(\hat{H}_{n,\mathbb{S}})$ of $\hat{H}_{n,\mathbb{S}}$ is the ring of symmetric Laurent polynomials in $X_1, \ldots, X_n$. That is,

$$Z(\hat{H}_{n,\mathbb{S}}) = \mathbb{S}\left[X_1^{\pm 1}, \ldots, X_n^{\pm 1}\right]^{S_n}.$$

We now turn to the Hecke algebra of type $G(m,1,n)$ and establish its basic properties. Recall that $t_1, \ldots, t_n$ are defined by

$$t_1 = a_1, \quad t_i = q^{-1} a_i t_{i-1} a_i \ (2 \le i \le n).$$

LEMMA 13.2. *We have the following equalities in* $\mathcal{H}_{n,\mathbb{S}}$.

(1) $a_i t_{i-1} = t_i a_i + (1-q)t_i \ (2 \le i \le n)$.
(2) $a_i t_i = t_{i-1} a_i + (q-1)t_i \ (2 \le i \le n)$.
(3) $a_i t_{i-1} t_i = t_i t_{i-1} a_i$, $a_i(t_{i-1} + t_i) = (t_{i-1} + t_i)a_i \ (2 \le i \le n)$.
(4) $a_i t_j = t_j a_i \ (j \ne i-1, i,\ 2 \le i \le n)$.
(5) $t_i t_j = t_j t_i \ (1 \le i, j \le n)$.
(6) *If* $0 \le e \le f$ *and* $2 \le i \le n$ *then*

$$a_i t_{i-1}^e t_i^f = t_{i-1}^f t_i^e a_i + (q-1)\left(\sum_{k=1}^{f-e} t_{i-1}^{e+k-1} t_i^{f-k+1}\right).$$

(7) *If* $0 \le f \le e$ *and* $2 \le i \le n$ *then*

$$a_i t_{i-1}^e t_i^f = t_{i-1}^f t_i^e a_i + (1-q)\left(\sum_{k=1}^{e-f} t_{i-1}^{e-k} t_i^{f+k}\right).$$

PROOF. Parts (1) and (2) follow from

$$(1)'\ a_i t_{i-1} = t_i(q a_i^{-1}) = t_i(a_i + (1-q)),$$
$$(2)'\ a_i t_i = (q^{-1}a_i^2)t_{i-1}a_i = (1 + (q-1)q^{-1}a_i)t_{i-1}a_i.$$

(3) Obvious consequence of (1) and (2).
(4) We may prove this by considering the two cases $j \ge i+1$ and $j \le i-2$.
(5) We may assume that $j > i$. Since $a_1, \ldots, a_i$ commute with t_j by (4), this is also obvious.

(6)(7) We may assume $e = 0$ or $f = 0$ by (3) and (5). Then the statement can be proved by induction. □

COROLLARY 13.3. *Let $\mathbb{S}$ be a ring and assume that the parameters v_i are all invertible. Then the following map defines a surjective algebra homomorphism from $\hat{H}_{n,\mathbb{S}}$ to $\mathcal{H}_{n,\mathbb{S}}$.*

$$\hat{H}_{n,\mathbb{S}} \to \mathcal{H}_{n,\mathbb{S}} : \begin{cases} T_i \mapsto a_i & (2 \le i \le n) \\ X_i^{\pm 1} \mapsto t_i^{\pm 1} & (1 \le i \le n) \end{cases}$$

Its proof follows by checking the defining relations and is left to the reader. The next corollary is subtler than it first looks, since the commutative subalgebra generated by $t_1, \ldots, t_n$ has $\mathbb{S}$-rank greater than m^n in general.

COROLLARY 13.4. *Let $\mathbb{S}$ be a ring. Then $\mathcal{H}_{n,\mathbb{S}}$ is generated by the following elements as an $\mathbb{S}$-module.*

$$\{ a_w t_1^{e_1} \cdots t_n^{e_n} \mid 0 \le e_i < m, \ w \in S_n \}$$

PROOF. Let $\mathcal{H}'_{n,\mathbb{S}}$ be the $\mathbb{S}$-submodule generated by these elements: then,

$$\mathcal{H}'_{n,\mathbb{S}} = \sum_{0 \le e_i < m, \ w \in S_n} \mathbb{S} a_w t_1^{e_1} \cdots t_n^{e_n}.$$

Since $\mathcal{H}'_{n,\mathbb{S}}$ contains 1, it is enough to show $\mathcal{H}'_{n,\mathbb{S}} a_i \subset \mathcal{H}'_{n,\mathbb{S}}$. But this is a consequence of Lemma 13.2(4),(6),(7). Note that (6) and (7) may be viewed as formulas for rewriting $t_{i-1}^e t_i^f a_i$ in terms of the basis above. □

13.2. Semi-normal representations

Let $\lambda = (\lambda^{(m)}, \ldots, \lambda^{(1)})$ be a (multi-)Young diagram as in Chapter 10. We denote by $\mathcal{ST}_\lambda$ the set of standard tableaux $\mathbf{t} = (\mathbf{t}^{(m)}, \ldots, \mathbf{t}^{(1)})$ of shape λ. Recall that $\lambda \vdash n$ means that the number of nodes in λ is n.

To consider the semisimplicity criterion of the Hecke algebra of type $G(m,1,n)$, we introduce the following parameter condition. Let $\mathbb{S}^\times$ be the set of units in $\mathbb{S}$.

DEFINITION 13.5. We say that the parameters of $\mathcal{H}_{n,\mathbb{S}}$ satisfy the **separation condition** if

$$\prod_{i=1}^{n} (1 + q + \cdots + q^{i-1}) \prod_{\substack{i<j \\ -n<d<n}} (q^d v_i - v_j) \in \mathbb{S}^\times.$$

The following theorem is a natural generalization of Hoefsmit's result on Hecke algebras of type A and type B; it will be used in subsequent sections. Recall that if $\boxed{\mathrm{i}}$ lies in the a th row and the b th column of $\lambda^{(c)}$, we denote a, b, c by row(i), col(i) and comp(i) respectively.

THEOREM 13.6. (1) *Let $\mathbb{K}$ be a field. Let $V_{\mathbb{K}}^\lambda$ be the $\mathbb{K}$-vector space with basis indexed by $\mathcal{ST}_\lambda$ and write*

$$V_{\mathbb{K}}^\lambda = \sum_{\mathbf{t} \in \mathcal{ST}_\lambda} \mathbb{K}\mathbf{t}.$$

Assume that for all $\mathbf{t} \in \mathcal{ST}_\lambda$, $v_{c_i} - q^{k_i} v_{c_{i-1}}$ in (13.1) below is invertible for all i. Then $V_{\mathbb{K}}^\lambda$ is the $\mathcal{H}_{n,\mathbb{K}}$-module with basis $\mathcal{ST}_\lambda$ and $\mathcal{H}_{n,\mathbb{K}}$-action given by:

– $a_1\mathbf{t} = v_{\mathrm{comp}(1)}\mathbf{t}$.
– *If* $i \geq 2$ *and two nodes* $\boxed{i\text{-}1}$ *and* $\boxed{i}$ *are adjacent, then*

$$a_i\mathbf{t} = \begin{cases} q\mathbf{t} & \text{(if they are in the same row)}, \\ -\mathbf{t} & \text{(if they are in the same column)}. \end{cases}$$

– *If* $i \geq 2$ *and two nodes* $\boxed{i\text{-}1}$ *and* $\boxed{i}$ *are not adjacent, then*

$$a_i(\mathbf{t}, \mathbf{t}') = (\mathbf{t}, \mathbf{t}')M_i(\mathbf{t}, \mathbf{t}'),$$

where $\mathbf{t}'$ *is the standard tableau obtained from* $\mathbf{t}$ *by exchanging* $\boxed{i\text{-}1}$ *and* $\boxed{i}$, *and* $M_i(\mathbf{t}, \mathbf{t}')$ *is the matrix given by*

$$\frac{1}{v_{c_i} - q^{k_i}v_{c_{i-1}}} \begin{pmatrix} (q-1)v_{c_i} & v_{c_i} - q^{k_i+1}v_{c_{i-1}} \\ qv_{c_i} - q^{k_i}v_{c_{i-1}} & (1-q)v_{c_i} \end{pmatrix}, \tag{13.1}$$

where $k_i = -(\mathrm{row}(i) - \mathrm{row}(i-1)) + (\mathrm{col}(i) - \mathrm{col}(i-1))$ *and* $c_i = \mathrm{comp}(i)$.

(2) *Let* $\mathrm{Res}(V_{\mathbb{K}}^{\lambda})$ *be the* $\mathcal{H}_{n-1,\mathbb{K}}$*-module obtained from* $V_{\mathbb{K}}^{\lambda}$ *by restriction. Then*

$$\mathrm{Res}(V_{\mathbb{K}}^{\lambda}) \simeq \bigoplus_{\mu:|\lambda/\mu|=1} V_{\mathbb{K}}^{\mu}.$$

Proof. (1) It is enough to check the defining relations of $a_1, \ldots, a_n$. This is an easy but long computation; see [**H-AK**].
(2) Restriction is given by ignoring the $\boxed{n}$ in each standard tableau; hence the restriction rule follows by inspection. □

If the assumption of Theorem 13.6 that for all $\mathbf{t} \in \mathcal{ST}_{\lambda}$, $v_{c_i} - q^{k_i}v_{c_{i-1}}$ is invertible for all i is satisfied, we call $\mathbb{K}$ **λ-separated**. Observe that the separation condition implies that $\mathbb{K}$ is λ-separated for all λ.

Definition 13.7. Representations given by the $\mathcal{H}_{n,\mathbb{K}}$-modules $V_{\mathbb{K}}^{\lambda}$ introduced in Theorem 13.6 are called **semi-normal** representations.

Corollary 13.8. *Assume that* $\mathbb{K}$ *is* λ*-separated. Then the basis elements* $\{\mathbf{t}|\mathbf{t} \in \mathcal{ST}_{\lambda}\}$ *of* $V_{\mathbb{K}}^{\lambda}$ *are simultaneous eigenvectors for* $t_1, \ldots, t_n$. *Furthermore, the simultaneous eigenvalues are given by*

$$t_i\mathbf{t} = \varphi_{\mathbf{t}}(t_i)\mathbf{t}, \quad \varphi_{\mathbf{t}}(t_i) = q^{-\mathrm{row}(i)+\mathrm{col}(i)}v_{\mathrm{comp}(i)}.$$

Proof. This is proved by induction on i. □

Corollary 13.9. *Let* $\mathbb{K}$ *be a field, and assume* $q \neq 1$ *and the separation condition. Then* $\{V_{\mathbb{K}}^{\lambda}\}_{\lambda\vdash n}$ *is a complete set of irreducible* $\mathcal{H}_{n,\mathbb{K}}$*-modules. Consequently,* $\mathcal{H}_{n,\mathbb{K}}$ *is a* $m^n n!$*-dimensional split semisimple* $\mathbb{K}$*-algebra.*

Proof. As the separation condition is imposed on $\mathbb{K}$, all $V_{\mathbb{K}}^{\lambda}$ are defined. First we prove that each $V_{\mathbb{K}}^{\lambda}$ is absolutely irreducible. Let $\mathbb{K}$ be algebraically closed and let W be a non-zero $\mathcal{H}_{n,\mathbb{K}}$-submodule of $V_{\mathbb{K}}^{\lambda}$. As in Corollary 13.8 define $\varphi_{\mathbf{t}}$ for each standard tableau $\mathbf{t} \in \mathcal{ST}_{\lambda}$; that is,

$$\varphi_{\mathbf{t}} : t_i \mapsto q^{-\mathrm{row}(i)+\mathrm{col}(i)}v_{\mathrm{comp}(i)}.$$

Then $\{\varphi_{\mathbf{t}}|\mathbf{t} \in \mathcal{ST}_{\lambda}\}$ is a complete set of simultaneous eigenvalues of $t_1, \ldots, t_n$ on $V_{\mathbb{K}}^{\lambda}$, and the separation condition implies that the $\varphi_{\mathbf{t}}$ are all distinct; $\varphi_{\mathbf{t}} = \varphi_{\mathbf{s}}$ implies $\mathbf{t} = \mathbf{s}$. To prove this, observe that if $i \neq j$ then $v_iq^k \neq v_j$, for $-n < k < n$, so that comp(i) takes the same value on $\mathbf{s}$ and $\mathbf{t}$ for each i. Further, no two of $1, q, \ldots, q^{n-1}$ are the same since $(q-1)(1+\cdots+q^k) \neq 0$ for $0 \leq k < n$, which implies that $-\mathrm{row}(i)+\mathrm{col}(i)$ takes the same value on $\mathbf{s}$ and $\mathbf{t}$ for each i. Assume that the (multi-)Young diagram formed by $\boxed{1}, \ldots, \boxed{i\text{-}1}$ are the same in both tableaux. Since $\boxed{i}$ is an addable node (for this (multi-)Young diagram) of the same residue, it follows that $\boxed{i}$ also occupies the same position in $\mathbf{s}$ and $\mathbf{t}$. Hence $\mathbf{s} = \mathbf{t}$ by induction.

A consequence of the claim above is that if $\sum c_{\mathbf{t}}\mathbf{t} \in W$ and $c_{\mathbf{t}} \neq 0$, then we have $\mathbf{t} \in W$: let $\mathcal{T}$ be the $\mathbb{K}$-subalgebra of $\mathcal{H}_{n,\mathbb{K}}$ generated by $\mathbf{t}_1, \ldots, \mathbf{t}_n$. Corollary 13.8 implies that $V_{\mathbb{K}}^{\lambda}$ is a semisimple $\mathcal{T}$-module and the claim says that $V_{\mathbb{K}}^{\lambda}$ is multiplicity-free as a $\mathcal{T}$-module. Hence the projection onto $\mathbb{K}\mathbf{t}$ is the image of an idempotent in $\mathcal{T}$. Thus applying this idempotent to $\sum c_{\mathbf{t}}\mathbf{t} \in W$ we get $\mathbf{t} \in W$.

Recalling that $W \neq 0$, we may assume that a standard tableau $\mathbf{t}$ lies in W. We want to prove that all other $\mathbf{s}$ belong to W. To prove this, it suffices to show that there is an a_w such that $\mathbf{s}$ appears in $a_w\mathbf{t}$. If we consider $w \in S_n$ such that $w\mathbf{t} = \mathbf{s}$, and choose w with $l(w)$ minimal, then w satisfies the requirement. Therefore, we have $W = V_{\mathbb{K}}^{\lambda}$, and $V_{\mathbb{K}}^{\lambda}$ is an absolutely irreducible $\mathcal{H}_{n,\mathbb{K}}$-module for any field $\mathbb{K}$ which satisfies the separation condition.

Next we prove that $\lambda \neq \mu$ implies $V_{\mathbb{K}}^{\lambda} \not\simeq V_{\mathbb{K}}^{\mu}$. Assume that the statement is true for those (multi-)Young diagrams with less than n nodes. Assume further that $V_{\mathbb{K}}^{\lambda} \simeq V_{\mathbb{K}}^{\mu}$ holds for λ and μ satisfying $|\lambda| = |\mu| = n \geq 2$. We show that we get a contradiction if $\lambda \neq \mu$. First of all, Theorem 13.6(2) implies that

$$\bigoplus_{\nu:|\lambda/\nu|=1} V_{\mathbb{K}}^{\nu} \simeq \bigoplus_{\nu':|\mu/\nu'|=1} V_{\mathbb{K}}^{\nu'}. \tag{13.2}$$

By the induction hypothesis, $V_{\mathbb{K}}^{\nu} \simeq V_{\mathbb{K}}^{\nu'}$ implies $\nu = \nu'$. Thus, there exists ν such that V^{ν} appears on both sides. Since we have assumed $\lambda \neq \mu$, we must have $\nu = \lambda \cap \mu$. Let $x \neq y$ be nodes such that $\lambda = \nu \cup \{x\}, \mu = \nu \cup \{y\}$. If μ has more than one removable node, one of these is different from y. Let z be such a node and consider $\nu' = \mu \setminus \{z\}$. Then $V_{\mathbb{K}}^{\nu'}$ appears on the right hand side of (13.2), but does not appear on the left hand side since $\mu \setminus \{z\}$ contains a node y which is not a node of λ. This is absurd, so we may conclude that $\mu^{(i_0)}$ is a rectangle for some i_0 and all other components are $\mu^{(i)} = \emptyset$. By the same argument, λ has the same form. This is impossible unless $n = 1$ since $\nu = \lambda \cap \mu$ implies that λ is obtained from μ by moving the unique removable node of $\mu^{(i_0)}$.

Finally, we prove that $\{V_{\mathbb{K}}^{\lambda}\}_{\lambda \vdash n}$ is a complete set of irreducible $\mathcal{H}_{n,\mathbb{K}}$-modules. Let $f^{\lambda^{(i)}}$ be the number of standard tableaux of shape $\lambda^{(i)}$. Since the Robinson-Schensted correspondence implies the combinatorial identity

$$\sum_{\lambda^{(i)} \vdash n_i} \left(f^{\lambda^{(i)}}\right)^2 = n_i!,$$

we have

$$\dim V_{\mathbb{K}}^{\lambda} = \frac{n!}{n_1! \cdots n_m!} f^{\lambda^{(1)}} \cdots f^{\lambda^{(m)}}$$

where $n_i = |\lambda^{(i)}|$ $(1 \leq i \leq m)$. Using this and the inequality

$$\dim \mathcal{H}_{n,\mathbb{K}} \geq \dim\left(\mathcal{H}_{n,\mathbb{K}}/\mathrm{rad}\,\mathcal{H}_{n,\mathbb{K}}\right) \geq \sum_{\lambda \vdash n} (\dim V_{\mathbb{K}}^{\lambda})^2,$$

we obtain $\dim\mathcal{H}_{n,\mathbb{K}} \geq m^n n!$. Since the opposite inequality is proved in Corollary 13.4, we may conclude that $\mathcal{H}_{n,\mathbb{K}}$ is a semisimple algebra and that the $V_{\mathbb{K}}^{\lambda}$ form a comlete set of irreducible $\mathcal{H}_{n,\mathbb{K}}$-modules. Since the $V_{\mathbb{K}}^{\lambda}$ are absolutely irreducible, we also find that $\mathcal{H}_{n,\mathbb{K}}$ is a split semisimple algebra. □

Recall that we say that $V_{\mathbb{K}}^{\lambda}$ is λ-separated if the denominators $v_{c_i} - q^{k_i}v_{c_{i-1}}$ of $M_i(\mathbf{t},\mathbf{t}')$ are all non-zero.

PROPOSITION 13.10. *Let $\mathbb{K}$ be a field and assume that $q \neq 1$. Then the following are equivalent.*

(1) *The separation condition is satisfied.*
(2) *$\mathbb{K}$ is λ-separated for all $\lambda \vdash n$.*
(3) *$\mathcal{H}_{n,\mathbb{K}}$ is a split semisimple $\mathbb{K}$-algebra.*
(4) *$\mathcal{H}_{n,\mathbb{K}}$ is a semisimple $\mathbb{K}$-algebra.*
(5) *$\mathbb{K}$ is λ-separated for all λ and the $V_{\mathbb{K}}^{\lambda}$ form a complete set of irreducible $\mathcal{H}_{n,\mathbb{K}}$-modules.*

PROOF. (1)⇔(2) Obvious.
(1)⇒(3)(5) Corollary 13.9.
(5)⇒(2) Obvious.
(3)⇒(4) Obvious.
(4)⇒(1) It is enough to show that $\mathcal{H}_{n,\mathbb{K}}$ is not semisimple in the following two cases.

$$\begin{cases} m=2;\ v_1=1, v_2=q^d\ (-n<d<n) \\ m=1;\ v_1=1, 1+\cdots+q^{k-1}=0\ (2\leq k<n) \end{cases}$$

To do this, we compute the central idempotents for the one dimensional representations

$$\begin{cases} a_1 \mapsto q^d,\ a_i \mapsto -1\ (i\geq 2) \\ a_1 \mapsto 1,\ a_2 \mapsto q\ (i \geq 2) \end{cases}$$

respectively. Then we get that candidates for these idempotents square to 0; see [**H-A**]. This means that these modules are not projective as $\mathcal{H}_{n,\mathbb{K}}$-modules. Hence, $\mathcal{H}_{n,\mathbb{K}}$ is not semisimple in these cases. □

PROPOSITION 13.11. *Let $\mathbb{S}$ be a ring. Then the following set is an $\mathbb{S}$-free basis of $\mathcal{H}_{n,\mathbb{S}}$.*

$$\{\, a_w t_1^{e_1}\cdots t_n^{e_n} \mid 0 \leq e_i < m,\ w \in S_n \,\} \tag{13.3}$$

PROOF. In the case where the parameters $\mathbf{q}, \mathbf{v}_1, \ldots, \mathbf{v}_m$ of $\mathcal{H}_{n,\mathbb{S}}$ are indeterminates and $\mathbb{S}$ is the field $\mathbb{Q}(\mathbf{q}, \mathbf{v}_1, \ldots, \mathbf{v}_m)$, this follows from Corollary 13.4 and Corollary 13.9.

Take the $\mathbb{Z}$-subalgebra $\mathbb{A} = \mathbb{Z}[\mathbf{q}^{\pm 1}, \mathbf{v}_1, \ldots, \mathbf{v}_m]$ of $\mathbb{Q}(\mathbf{q}, \mathbf{v}_1, \ldots, \mathbf{v}_m)$, and consider the case where $\mathbb{S} = \mathbb{A}$. We have the natural algebra homomorphism

$$\begin{array}{ccc} \mathcal{H}_{n,\mathbb{A}} & \longrightarrow & \mathcal{H}_{n,\mathbb{Q}(\mathbf{q},\mathbf{v}_1,\ldots,\mathbf{v}_m)} \\ a_i & \mapsto & a_i. \end{array}$$

By Corollary 13.4, the set given in (13.3) generates $\mathcal{H}_{n,\mathbb{A}}$ as an $\mathbb{A}$-module, and the images of these elements are linearly independent vectors of a $\mathbb{K}$-vector space. Therefore, the set (13.3) is an $\mathbb{A}$-free basis of $\mathcal{H}_{n,\mathbb{A}}$.

Finally, let $\mathbb{S}$ be an arbitrary ring. Then there exists an algebra homomorphism

$$\begin{array}{ccc} \mathcal{H}_{n,\mathbb{S}} & \longrightarrow & \mathcal{H}_{n,\mathbb{A}} \otimes \mathbb{S} \\ a_i & \mapsto & a_i \otimes 1 \end{array}$$

where the parameters of $\mathbb{A}$ map to those of $\mathbb{S}$: $\mathbf{q} \mapsto q, \mathbf{v}_i \mapsto v_i$ $(1 \le i \le m)$. It is obvious from the definitions that this homomorphism is an isomorphism. □

COROLLARY 13.12. *Let $\mathbb{S}$ be a ring and consider $\mathcal{H}_{n,\mathbb{S}}$ as a right $\mathcal{H}_{n-1,\mathbb{S}}$-module. Then $\mathcal{H}_{n,\mathbb{S}}$ is a free $\mathcal{H}_{n-1,\mathbb{S}}$-module.*

PROOF. Since a_w $(w \in S_{n-1})$ and t_n commute, Corollary 13.11 implies that

$$\mathcal{H}_{n,\mathbb{S}} = \bigoplus_{0 \le e < m} \bigoplus_{w \in S_n/S_{n-1}} a_w t_n^e \mathcal{H}_{n-1,\mathbb{S}},$$

where $w \in S_n/S_{n-1}$ means that w runs over the distinguished coset representatives of S_{n-1} in S_n. (It is easy to check that each coset contains a unique element of minimal length, which is called the distinguished coset representative of the coset.) The corollary follows. □

COROLLARY 13.13. *Let $\mathbb{S}$ be a ring. Then the functors*

$$\begin{array}{rccc} \mathrm{Ind}: & \mathcal{H}_{n,\mathbb{S}} - mod & \longrightarrow & \mathcal{H}_{n+1,\mathbb{S}} - mod \\ & V & \mapsto & \mathcal{H}_{n+1,\mathbb{S}} \underset{\mathcal{H}_{n,\mathbb{S}}}{\otimes} V \end{array}$$

$$\begin{array}{rccc} \mathrm{Res}: & \mathcal{H}_{n,\mathbb{S}} - mod & \longrightarrow & \mathcal{H}_{n-1,\mathbb{S}} - mod \\ & V & \mapsto & V \downarrow_{\mathcal{H}_{n-1,\mathbb{S}}} \end{array}$$

are exact functors.

PROOF. This follows from Corollary 13.12. □

By Corollary 13.13, Ind and Res induce operators on the Grothendieck groups of the corresponding module categories.

DEFINITION 13.14. Define operators Ind and Res on the Grothendieck groups by

$$\begin{array}{rccc} \mathrm{Ind}: & K_0(\mathcal{H}_{n,\mathbb{S}} - mod) & \longrightarrow & K_0(\mathcal{H}_{n+1,\mathbb{S}} - mod) \\ & [V] & \mapsto & [\mathrm{Ind}(V)] \end{array}$$

$$\begin{array}{rccc} \mathrm{Res}: & K_0(\mathcal{H}_{n,\mathbb{S}} - mod) & \longrightarrow & K_0(\mathcal{H}_{n-1,\mathbb{S}} - mod) \\ & [V] & \mapsto & [\mathrm{Res}(V)] \end{array}$$

13.3. The decomposition map

DEFINITION 13.15. The triple $(\mathbb{K}, \mathbb{S}, \mathbb{F})$ is a **modular system with parameters** if we are given parameters

$$\mathbf{q} \in \mathbb{S}^\times, \mathbf{v}_1, \ldots, \mathbf{v}_m \in \mathbb{S}; \quad q \in \mathbb{F}^\times, v_1, \ldots, v_m \in \mathbb{F}$$

such that the following conditions are satisfied.

(1) $\mathbb{S}$ is a discrete valuation ring, whose unique maximal ideal is (π).
(2) $\mathbb{K}$ is the fraction field of $\mathbb{S}$.
(3) $\mathbb{F} = \mathbb{S}/(\pi)$,i.e. $\mathbb{F}$ is the residue field of $\mathbb{S}$.

(4) $\mathbf{q} \pmod{\pi} = q$, $\mathbf{v}_i \pmod{\pi} = v_i$ $(1 \le i \le m)$.

If we drop the parameters, we call this triple a modular system.

Whenever a modular system with parameters is given, it is accompanied by the Hecke algebras $\mathcal{H}_{n,\mathbb{K}}$ and $\mathcal{H}_{n,\mathbb{S}}$ of type $G(m, 1, n)$ associated with parameters $(\mathbf{q}, \mathbf{v}_1, \dots, \mathbf{v}_m)$ and $\mathcal{H}_{n,\mathbb{F}}$ associated with parameters $(q, v_1, \dots, v_m)$. As in the proof of Proposition 13.11, we have $\mathcal{H}_{n,\mathbb{K}} = \mathcal{H}_{n,\mathbb{S}} \otimes_{\mathbb{S}} \mathbb{K}$ and $\mathcal{H}_{n,\mathbb{F}} = \mathcal{H}_{n,\mathbb{S}} \otimes_{\mathbb{S}} \mathbb{F}$.

The following two lemmas are well-known.

LEMMA 13.16. *Let $(\mathbb{K}, \mathbb{S}, \mathbb{F})$ be a modular system with parameters, V a $\mathcal{H}_{n,\mathbb{K}}$-module. Let L be a $\mathcal{H}_{n,\mathbb{S}}$-submodule of V which is an $\mathbb{S}$-lattice of full rank. Then $[L \otimes \mathbb{F}] \in K_0(\mathcal{H}_{n,\mathbb{F}} - mod)$ is determined by V and does not depend on the choice of L.*

PROOF. Note that such a lattice L exists since $\mathbb{S}$-submodules of V are torsion-free modules of the discrete valuation ring $\mathbb{S}$, and thus free $\mathbb{S}$-modules.

If L and L' are $\mathcal{H}_{n,\mathbb{S}}$-submodules which are full rank $\mathbb{S}$-lattices of V, then so is $L + L'$. Thus to prove the lemma, we may assume that $L \subset L'$. We may further assume that L is a maximal $\mathcal{H}_{n,\mathbb{S}}$-submodule of L': as

$$\pi^N L' \subset L \subset L'$$

for sufficiently large N and $L'/\pi^N L'$ is finitely generated as a module over $\mathbb{S}/(\pi^N)$, we can insert a finite number of submodules $L_1, L_2, \dots$ between L and L' such that L_i is a maximal $\mathcal{H}_{n,\mathbb{S}}$-submodule of L_{i+1}.

First we prove that $\pi L' \subset L$. If $\pi L' \not\subset L$ then, since L is a maximal $\mathcal{H}_{n,\mathbb{S}}$-submodule of L', we get $L' = L + \pi L'$. Thus we have $L = L'$ by Nakayama's Lemma, which contradicts the choice of L and L'. Hence we have $\pi L' \subset L$. Using $\pi L \subset \pi L' \subset L \subset L'$, we get

$$[L/\pi L] = [L/\pi L'] + [\pi L'/\pi L] = [L/\pi L'] + [L'/L] = [L'/\pi L'].$$

In particular, we have $[L \otimes \mathbb{F}] = [L' \otimes \mathbb{F}]$. □

LEMMA 13.17. *Let $(\mathbb{K}, \mathbb{S}, \mathbb{F})$ be a modular system with parameters. If*

$$0 \to V_1 \to V_2 \to V_3 \to 0$$

is an exact sequence of $\mathcal{H}_{n,\mathbb{K}}$-modules, then we may choose $\mathcal{H}_{n,\mathbb{S}}$-submodules L_i of V_i for $i = 1, 2, 3$ which are $\mathbb{S}$-lattices of full rank so that

$$0 \to L_1 \to L_2 \to L_3 \to 0$$

is an exact sequence of $\mathcal{H}_{n,\mathbb{S}}$-modules. In particular,

$$0 \to L_1 \otimes \mathbb{F} \to L_2 \otimes \mathbb{F} \to L_3 \otimes \mathbb{F} \to 0$$

is an exact sequence of $\mathcal{H}_{n,\mathbb{F}}$-modules.

PROOF. Choose an $\mathcal{H}_{n,\mathbb{S}}$-submodule L_2 of V_2 which is an $\mathbb{S}$-lattice of full rank. We identify V_1 with a submodule of V_2 and set $L_1 := L_2 \cap V_1$. Let L_3 be the full rank $\mathbb{S}$-lattice of V_3 obtained as the image of L_2. Then it is easy to see that the sequence

$$0 \to L_1 \to L_2 \to L_3 \to 0 \tag{13.4}$$

is exact. In particular, $\operatorname{rank} L_i = \dim V_i$ for $i = 2, 3$; this implies that L_1 is a full rank $\mathbb{S}$-lattice of V_1. Hence, the first half is proved.

Since L_3 is a free $\mathbb{S}$-module, the exact sequence of (13.4) splits. Hence, we obtain an exact sequence after tensoring with $\mathbb{F}$. Since the parameters correspond, the resulting sequence is an exact sequence of $\mathcal{H}_{n,\mathbb{F}}$-modules. □

By virtue of Lemma 13.16 and Lemma 13.17, we may define decomposition maps.

DEFINITION 13.18. Let $(\mathbb{K}, \mathbb{S}, \mathbb{F})$ be a modular system with parameters. Then the **decomposition map** is the map defined by

$$\begin{array}{rccc} d_{\mathbb{K},\mathbb{F}}: & K_0(\mathcal{H}_{n,\mathbb{K}} - mod) & \longrightarrow & K_0(\mathcal{H}_{n,\mathbb{F}} - mod) \\ & [V] & \mapsto & [L \otimes \mathbb{F}] \end{array}$$

where L is a full sublattice of V.

Note that we do not use any special properties of the Hecke algebra of type $G(m, 1, n)$ in the proof of Lemma 13.16 and Lemma 13.17. Thus, the same proof works for affine Hecke algebras. Thus, we may also define decomposition maps for affine Hecke algebras if we consider the module category of finite dimensional modules. In this case, the triple $(\mathbb{K}, \mathbb{S}, \mathbb{F})$ is a **modular system with parameters** if we are given parameters $\mathbf{q} \in \mathbb{S}^\times, q \in \mathbb{F}^\times$ and if the conditions (1),(2) and (3) of Definition 13.15 and (4') $q = \mathbf{q} \pmod{\pi}$ are satisfied.

The triple is accompanied by the affine Hecke algebras $\hat{H}_{n,\mathbb{K}}$ and $\hat{H}_{n,\mathbb{S}}$ associated with $\mathbf{q}$ and by the affine Hecke algebra $\hat{H}_{n,\mathbb{F}}$ associated with q.

LEMMA 13.19. *Let $(\mathbb{K}, \mathbb{S}, \mathbb{F})$ be a modular system with parameters. Then the map*

$$\begin{array}{rccc} d_{\mathbb{K},\mathbb{F}}: & K_0(\hat{H}_{n,\mathbb{K}} - mod) & \longrightarrow & K_0(\hat{H}_{n,\mathbb{F}} - mod) \\ & [V] & \mapsto & [L \otimes \mathbb{F}] \end{array}$$

is well-defined.

We call this map the decomposition map of affine Hecke algebras.

13.4. Specht module theory

To study the decomposition map of the Hecke algebras of type $G(m, 1, n)$, it is convenient to have matrix representations which are amenable to specializing parameters. Such matrix representations were constructed by Dipper, James and Mathas [**S-DJM2**]. Another construction was given by Graham and Lehrer [**S-GL**], but it has a different flavor; the construction of Dipper-James-Mathas is a more natural generalization of the classical Specht module theory. The best way to show that the simple modules obtained from the Specht module theory give a complete set of simple modules is to appeal to the general theory of cellular algebras developed by Graham and Lehrer [**S-GL**].

DEFINITION 13.20. Let Λ_n be the poset $\{\lambda = (\lambda^{(m)}, \ldots, \lambda^{(1)})\}_{\lambda \vdash n}$. The partial order of Λ_n, called the **dominance ordering**, is defined as follows.

$$\lambda \trianglerighteq \mu \Leftrightarrow \sum_{i=1}^{k-1} |\lambda^{(i)}| + \sum_{j=1}^{l} \lambda_j^{(k)} \geq \sum_{i=1}^{k-1} |\mu^{(i)}| + \sum_{j=1}^{l} \mu_j^{(k)} \quad (\forall k, l)$$

where $\lambda_1^{(k)} \geq \lambda_2^{(k)} \geq \cdots$, $\mu_1^{(k)} \geq \mu_2^{(k)} \geq \cdots$ are the parts of $\lambda^{(k)}$ and $\mu^{(k)}$.

THEOREM 13.21 (Dipper-James-Mathas). *Let $\mathbb{A}$ be the Laurent polynomial ring*

$$\mathbb{A} = \mathbb{Z}[\mathbf{q}^{\pm 1}, \mathbf{v}_1, \ldots, \mathbf{v}_m].$$

Then, for each $\lambda \in \Lambda_n$ there is an $\mathcal{H}_{n,\mathbb{A}}$-module $S_{\mathbb{A}}^{\lambda}$ with the following properties.

(1) *$S_{\mathbb{A}}^{\lambda}$ is free as an $\mathbb{A}$-module and it is equipped with an invariant symmetric bilinear form. By an invariant form, we mean $\langle a_i u, u' \rangle = \langle u, a_i u' \rangle$, for $1 \le i \le n$.*

(2) *$S_{\mathbb{A}}^{\lambda}$ has an $\mathbb{A}$-module filtration*

$$S_{\mathbb{A}}^{\lambda} = F(0) \supset F(1) \supset \cdots$$

such that these submodules $F(i)$ are stable under $t_1, \ldots, t_n$ and each successive quotient $F(i)/F(i+1)$ is a free $\mathbb{A}$-module of rank 1. Further, as a multiset, the set of the simultaneous eigenvalues of the action of $t_1, \ldots, t_n$ on the quotients $F(i)/F(i+1)$ is equal to $\{\varphi_{\mathbf{t}}\}_{\mathbf{t} \in ST_{\lambda}}$.

(3) *Let $\mathbb{K}$ be a field and assume that $\mathcal{H}_{n,\mathbb{K}}$ is semisimple. Then we have the following isomorphism of $\mathcal{H}_{n,\mathbb{K}}$-modules.*

$$S_{\mathbb{K}}^{\lambda} := S_{\mathbb{A}}^{\lambda} \otimes \mathbb{K} \simeq V_{\mathbb{K}}^{\lambda},$$

where $V_{\mathbb{K}}^{\lambda}$ is the module introduced in Theorem 13.6.

(4) *Let $\mathbb{F}$ be a field. If we set $S_{\mathbb{F}}^{\lambda} := S_{\mathbb{A}}^{\lambda} \otimes \mathbb{F}$ and $D_{\mathbb{F}}^{\lambda} := S_{\mathbb{F}}^{\lambda}/\mathrm{rad}_{\langle\ ,\ \rangle} S_{\mathbb{F}}^{\lambda}$, then $D_{\mathbb{F}}^{\lambda}$ is either 0 or an absolutely irreducible $\mathcal{H}_{n,\mathbb{F}}$-module.*

(5) *$\{D_{\mathbb{F}}^{\lambda} \neq 0\}$ is a complete set of simple $\mathcal{H}_{n,\mathbb{F}}$-modules.*

(6) *Assume that $D_{\mathbb{F}}^{\mu} \neq 0$. Then $d_{\lambda,\mu} := [S_{\mathbb{F}}^{\lambda} : D_{\mathbb{F}}^{\mu}] \neq 0$ only if $\mu \trianglelefteq \lambda$. Further, we have $[S_{\mathbb{F}}^{\mu} : D_{\mathbb{F}}^{\mu}] = 1$.*

We sketch the strategy used to prove this. We set $l_j = \sum_{k=1}^{j-1} |\lambda^{(k)}|$ for $\lambda = (\lambda^{(m)}, \ldots, \lambda^{(1)})$. The row stabilizer of λ is $S_{\lambda} = S_{\lambda^{(m)}} \times \cdots \times S_{\lambda^{(1)}}$. We define $m_{\lambda} \in \mathcal{H}_{n,\mathbb{A}}$ by

$$m_{\lambda} = \left(\prod_{j=2}^{m} \prod_{i=1}^{l_j} (t_i - v_j) \right) \left(\sum_{w \in S_{\lambda}} a_w \right).$$

Next we define $d(\mathbf{t}) \in S_n$ for a standard tableau $\mathbf{t}$. The canonical tableau $\mathbf{t}^{\lambda}$ is the standard tableau which is obtained by filling the nodes with $1, \ldots, n$ from right to left starting with the first row of $\lambda^{(1)}$ and ending with the last row of $\lambda^{(m)}$. Let $d(\mathbf{t})$ be the permutation which sends $\mathbf{t}^{\lambda}$ to $\mathbf{t}$, that is, $d(\mathbf{t})(i) = j$ if a node of λ has entry i in $\mathbf{t}^{\lambda}$ and j in $\mathbf{t}$.

Now we introduce elements $m_{\mathbf{s},\mathbf{t}}$ indexed by pairs $(\mathbf{s}, \mathbf{t})$ of standard tableaux of same shape as follows.

$$m_{\mathbf{s},\mathbf{t}} = a_{d(\mathbf{s})} m_{\lambda} a_{d(\mathbf{t})^{-1}}$$

Let $\overline{\mathcal{N}}^{\lambda}$ be the $\mathbb{A}$-submodule of $\mathcal{H}_{n,\mathbb{A}}$ generated by the $m_{\mathbf{s},\mathbf{t}}$ where $\mathbf{s}$ and $\mathbf{t}$ have shape μ for some $\mu \triangleright \lambda$. Similarly, let $\mathcal{N}^{\lambda}$ be the $\mathbb{A}$-submodule of $\mathcal{H}_{n,\mathbb{A}}$ generated by the $m_{\mathbf{s},\mathbf{t}}$ where $\mathbf{s}$ and $\mathbf{t}$ have shape μ for some $\mu \trianglerighteq \lambda$.

By computing $a_i m_{\mathbf{s},\mathbf{t}}$ and applying the anti-involution fixing a_i, we can prove that these $\mathbb{A}$-submodules are two-sided ideals of $\mathcal{H}_{n,\mathbb{A}}$ [**S-DJM2**, Proposition 3.22].

The module $S_{\mathbb{A}}^{\lambda}$ is the $\mathcal{H}_{n,\mathbb{A}}$-submodule of $\mathcal{N}^{\lambda}/\overline{\mathcal{N}}^{\lambda}$ generated by

$$z_{\lambda} := m_{\lambda} \pmod{\overline{\mathcal{N}}^{\lambda}}.$$

$S^\lambda_{\mathbb{A}}$ has $\{m_{\mathbf{st}^\lambda}\}_{\mathbf{s}\in ST_\lambda}$ as an $\mathbb{A}$-free basis and the invariant symmetric bilinear form is given by

$$z_\lambda a_{d(\mathbf{s})^{-1}} a_{d(\mathbf{t})} z_\lambda = \langle m_{\mathbf{st}^\lambda}, m_{\mathbf{tt}^\lambda}\rangle z_\lambda.$$

Hence we have (1).

(2) follows from Theorem 13.6 and Murphy's lemma [**cH-Mu**, Lemma 3.8]. (3) follows from (2). (4) is obtained by a standard argument. Parts (5) and (6) may be proved using the theory of cellular algebras.

DEFINITION 13.22. The module $S^\lambda_{\mathbb{F}}$ introduced in Theorem 13.21 is called the **Specht module**. The multiplicities $d_{\lambda,\mu}$ are called **decomposition numbers**.

Note that the decomposition numbers are determined by $\mathbb{F}$ and no modular system is involved in our definition.

COROLLARY 13.23. *Let $(\mathbb{K},\mathbb{S},\mathbb{F})$ be a modular system with parameters, $d_{\mathbb{K},\mathbb{F}}$ the decomposition map introduced in Definition 13.18. Then we have the following.*

(1) $d_{\mathbb{K},\mathbb{F}}([S^\lambda_{\mathbb{K}}]) = [S^\lambda_{\mathbb{F}}]$.
(2) $d_{\mathbb{K},\mathbb{F}}$ *is surjective.*
(3) *If $v_1,\dots,v_m$ are powers of q then the eigenvalues of $t_1,\dots,t_n$ are powers of q for any $\mathcal{H}_{n,\mathbb{F}}$-module.*
(4) *The decomposition map $d_{\mathbb{K},\mathbb{F}}$ of the Hecke algebra of type $G(m,1,n)$ does not depend on the choice of $\mathbb{S}$.*

PROOF. (1) Since $S^\lambda_{\mathbb{S}}\otimes\mathbb{K} = S^\lambda_{\mathbb{K}}$ and $S^\lambda_{\mathbb{S}}\otimes\mathbb{F} = S^\lambda_{\mathbb{F}}$, this is obvious.
(2) Let $\Lambda'_n := \{\,\lambda\in\Lambda_n \mid D^\lambda_{\mathbb{F}}\neq 0\,\}$. Then we have for $\lambda\in\Lambda'_n$ that

$$[S^\lambda_{\mathbb{F}}] = [D^\lambda_{\mathbb{F}}] + \sum_{\mu\lhd\lambda,\mu\in\Lambda'_n} d_{\lambda,\mu}[D^\mu_{\mathbb{F}}].$$

We order elements of Λ'_n linearly by using a total order which is compatible with the dominance order. Then the matrix $(d_{\lambda,\mu})_{\lambda,\mu\in\Lambda'_n}$ is triangular, and its diagonal entries are all 1. In particular, we may describe $[D^\lambda_{\mathbb{F}}]$ as a linear combination of $[S^\mu_{\mathbb{F}}]$ $(\mu\unlhd\lambda)$ with integral coefficients. Since all the $[S^\mu_{\mathbb{F}}]$ belong to the image of $d_{\mathbb{K},\mathbb{F}}$ by (1), we have the result.
(3) By (2), it is enough to prove the statement for $S^\lambda_{\mathbb{F}}$: this is already proved in Theorem 13.21(2).
(4) This follows from (1) and (2). □

We may apply Theorem 13.21 to prove the surjectivity of the decomposition map of the affine Hecke algebra.

COROLLARY 13.24. *Let $(\mathbb{K},\mathbb{S},\mathbb{F})$ be a modular system and assume that $\mathbb{F}$ is algebraically closed. Then the decomposition map*

$$d_{\mathbb{K},\mathbb{F}} : K_0(\hat{H}_{n,\mathbb{K}} - mod) \longrightarrow K_0(\hat{H}_{n,\mathbb{F}} - mod)$$

is surjective.

PROOF. It is enough to prove that any simple $\hat{H}_{n,\mathbb{F}}$-module is in the image of the decomposition map. Fix a simple $\hat{H}_{n,\mathbb{F}}$-module and denote the roots of the characteristic polynomial of X_1 on this module by $v_1,\dots,v_m$. Since $\mathbb{F}$ is algebraically closed, these parameters are in $\mathbb{F}$. We consider the Hecke algebra $\mathcal{H}_{n,\mathbb{F}}$ of type

$G(m,1,n)$ associated with $(q, v_1, \dots, v_m)$. Then we may view this simple $\hat{H}_{n,\mathbb{F}}$-module as an $\mathcal{H}_{n,\mathbb{F}}$-module by Corollary 13.3. We choose $\mathbf{v}_1, \dots, \mathbf{v}_m \in \mathbb{S}$ satisfying $\mathbf{v}_i \pmod{\pi} = v_i$. Then $(\mathbb{K}, \mathbb{S}, \mathbb{F})$ is a modular system with parameters. Since the decomposition map from the Hecke algebra $\mathcal{H}_{n,\mathbb{K}}$ associated with $(\mathbf{q}, \mathbf{v}_1, \dots, \mathbf{v}_m)$ to the Hecke algebra $\mathcal{H}_{n,\mathbb{F}}$ is surjective by Corollary 13.23(2), this simple $\hat{H}_{n,\mathbb{F}}$-module may be described by a linear combination of Specht modules in the Grothendieck group. Since the Specht modules $S^\lambda_\mathbb{K}$ may be viewed as $\hat{H}_{n,\mathbb{K}}$-modules by Corollary 13.3, we have the result. □

In later arguments, we do not need the whole $\hat{H}_{n,\mathbb{F}} - mod$ but only a certain full subcategory. Thus we introduce this subcategory here, and study the decomposition map restricted to this subcategory.

DEFINITION 13.25. Let $\mathcal{C}_{n,\mathbb{F}}$ be the full subcategory of $\hat{H}_{n,\mathbb{F}} - mod$ whose objects are $\hat{H}_{n,\mathbb{F}}$-modules on which the eigenvalues of $X_1, \dots, X_n$ are powers of q.

COROLLARY 13.26. *Let $(\mathbb{K}, \mathbb{S}, \mathbb{F})$ be a modular system with parameters with $\mathbb{F}$ being algebraically closed. Then the decomposition map*

$$d_{\mathbb{K},\mathbb{F}} : K_0(\mathcal{C}_{n,\mathbb{K}}) \longrightarrow K_0(\mathcal{C}_{n,\mathbb{F}})$$

is well-defined and surjective.

PROOF. If $[V] \in K_0(\mathcal{C}_{n,\mathbb{K}})$ then $d_{\mathbb{K},\mathbb{F}}([V]) \in K_0(\mathcal{C}_{n,\mathbb{F}})$ by definition. Thus the map is well-defined. To prove surjectivity, we fix a simple $\mathcal{H}_{n,\mathbb{F}}$-module which belongs to $\mathcal{C}_{n,\mathbb{F}}$ and follow the proof of Corollary 13.24. Since we may choose $v_1, \dots, v_m$ to be powers of q here, we may also choose $\mathbf{v}_1, \dots, \mathbf{v}_m$ to be powers of $\mathbf{q}$. Thus Corollary 13.23(2) implies that this simple $\mathcal{H}_{n,\mathbb{F}}$-module may be described as a linear combination of Specht modules and Corollary 13.23(3) implies that these Specht modules belong to $\mathcal{C}_{n,\mathbb{K}}$ as $\hat{H}_{n,\mathbb{K}}$-modules. Hence $d_{\mathbb{K},\mathbb{F}}$ is surjective. □

The following Lemma 13.27 also follows immediately from Corollary 13.23(3).

LEMMA 13.27. *If the parameters $v_1, \dots, v_m$ of the Hecke algebra $\mathcal{H}_{n,\mathbb{F}}$ of type $G(m,1,n)$ are powers of q then $\mathcal{H}_{n,\mathbb{F}} - mod$ is a full subcategory of $\mathcal{C}_{n,\mathbb{F}}$.*

We now define certain modules of the affine Hecke algebras. In Theorem 14.43 below, we will identify these modules with the images of the "standard modules" in the Grothedieck group of the module category of the affine Hecke algebra.

DEFINITION 13.28. Let $\mathbb{F}$ be a field and assume that we are given elements $x_1, \dots, x_l \in \mathbb{F}^\times$ and natural numbers $n_1, \dots, n_l$ whose sum is n. We denote by $\hat{H}_{n_1,\mathbb{F}} \otimes \cdots \otimes \hat{H}_{n_l,\mathbb{F}}$ the $\mathbb{F}$-subalgebra of $\hat{H}_{n,\mathbb{F}}$ generated by

$$T_i \ (i \neq n_1+1, n_1+n_2+1, \dots), \ X_1, \dots, X_n.$$

Let $M_{(x_1,\dots,x_l;n_1,\dots,n_l)}$ be the $\hat{H}_{n,\mathbb{F}}$-module obtained by inducing the one dimensional module of $\hat{H}_{n_1,\mathbb{F}} \otimes \cdots \otimes \hat{H}_{n_l,\mathbb{F}}$ which affords the representation

$$X_1 \mapsto x_1, \ X_2 \mapsto x_1 q, \dots, \ X_{n_1} \mapsto x_1 q^{n_1 - 1},$$
$$X_{n_1+1} \mapsto x_2, \ X_{n_1+2} \mapsto x_2 q, \dots,$$
$$\dots\dots.$$

LEMMA 13.29. *Suppose that $x_1, \ldots, x_l$ are powers of q. Then*

$$M_{(x_1,\ldots,x_l;n_1,\ldots,n_l)} \in \mathcal{C}_{n,\mathbb{F}}.$$

PROOF. Let $\mathbb{F}u$ be the one dimensional module of $\hat{H}_{n_1,\mathbb{F}} \otimes \cdots \otimes \hat{H}_{n_l,\mathbb{F}}$ which defines $M_{(x_1,\ldots,x_l;n_1,\ldots,n_l)}$. We write $X_\lambda u = \varphi(X_\lambda)u$. Define a subspace M_j of $M_{(x_1,\ldots,x_l;n_1,\ldots,n_l)}$ by

$$M_j = \sum_{w \in S_n : l(w) \leq j} \mathbb{F}\, T_w u.$$

Then we can prove for $w \neq 1$ that

$$X_\lambda T_w u \in \varphi(X_{w^{-1}\lambda}) T_w u + M_{l(w)-1}$$

by induction on $l(w)$. Hence M_j is stable under the action of $X_1, \ldots, X_n$ and X_i acts on M_j/M_{j-1} by powers of q. The result follows. □

13.5. A theorem of Morita type

The following Morita equivalence theorem of Dipper-Mathas reduces the study of the module category $\mathcal{H}_{n,\mathbb{F}} - mod$ to the case where the parameters $v_1, \ldots, v_m$ are powers of q.

THEOREM 13.30 (Dipper-Mathas). *Let $\mathbb{F}$ be a field and let $\mathcal{H}_{n,\mathbb{F}}(q;S)$ be the Hecke algebra of type $G(m,1,n)$ with parameters q and $S = \{v_1, \ldots, v_m\}$. We partition S into equivalence classes by declaring that two parameters v_i and v_j are in the same class if $q^{\mathbb{Z}} v_i = q^{\mathbb{Z}} v_j$, that is, if $q^z v_i = v_j$ for some $z \in \mathbb{Z}$. Denote this decomposition into equivalence classes by $S = \sqcup_{k=1}^a S_k$. Then we have the following (Morita) equivalence of categories.*

$$\mathcal{H}_{n,\mathbb{F}}(q;S) - mod \simeq \bigoplus_{(n_1,\ldots,n_a) \models n} \bigotimes_{k=1}^{a} \mathcal{H}_{n_k,\mathbb{F}}(q;S_k) - mod$$

where $(n_1, \ldots, n_a) \models n$ means $\sum_{k=1}^a n_k = n$.

Let $\mathbb{F}$ be a field and let A and B be finite dimensional $\mathbb{F}$-algebras. We say that two module categories $A - mod$ and $B - mod$ are **Morita-equivalent** if there exist covariant functors $F : A - mod \to B - mod$ and $G : B - mod \to A - mod$ which induce $\mathbb{F}$-linear maps $\mathrm{Hom}_A(M,N) \to \mathrm{Hom}_B(F(M), F(N))$ such that there are natural transformations $F \circ G \simeq 1_{B-mod}$ and $G \circ F \simeq 1_{A-mod}$. We write $A - mod \simeq B - mod$ if they are Morita equivalent.

A right A-module P is called a **progenerator** of $mod - A$ if P is a projective right A-module and any indecomposable projective right A-module appears in P as a direct summand.

The following lemma is standard.

LEMMA 13.31 (Morita). *Let $\mathbb{F}$ be a field, A a finite dimensional $\mathbb{F}$-algebra, P a progenerator of $mod - A$. Set $B = \mathrm{End}_A(P)$. Then the functors $F(-) = P \otimes -$ and $G(-) = \mathrm{Hom}_B(P, -)$ give a Morita equivalence of categories $A - mod \simeq B - mod$.*

In fact, Morita's famous theorem says that every Morita equivalence occurs in this way.

As this lemma suggests, Theorem 13.30 is proved by constructing a progenerator P of $mod - \mathcal{H}_{n,\mathbb{F}}(q; S)$ and showing that there is an isomorphism of algebras

$$\text{End}_A(P) \simeq \bigoplus_{(n_1,\ldots,n_a)\models n} \bigotimes_{k=1}^{a} \mathcal{H}_{n_k,\mathbb{F}}(q; S_k). \tag{13.5}$$

For the proof of Theorem 13.30 see [**cH-DM**].

By virtue of Theorem 13.30, we may assume that parameters v_i are powers of q to study the representations of the Hecke algebra of type $G(m, 1, n)$. (Renormalize a_1 if necessary.) Hence until the end of Chapter 14, we shall assume the following parameter condition.

[Parameter condition]: $q = \sqrt[r]{1}$ $(r \geq 2)$, $v_i = q^{\gamma_i}$ $(\gamma_i \in \mathbb{Z}/r\mathbb{Z})$

13.6. Refined induction and restriction functors

LEMMA 13.32. *The following formulas hold.*

$$\text{Ind}([S_{\mathbb{F}}^{\lambda}]) = \sum_{\mu:|\mu/\lambda|=1} [S_{\mathbb{F}}^{\mu}], \quad \text{Res}([S_{\mathbb{F}}^{\lambda}]) = \sum_{\mu:|\lambda/\mu|=1} [S_{\mathbb{F}}^{\mu}]$$

PROOF. We choose a modular system with parameters $(\mathbb{K}, \mathbb{S}, \mathbb{F})$ so that $\mathcal{H}_{n,\mathbb{K}}$ is semisimple. This is possible. In fact, if we take an indeterminate $\mathbf{t}$ and set

$$\mathbb{K} = \mathbb{F}(\mathbf{t}), \ \mathbb{S} = \mathbb{F}[\mathbf{t}]_{(\mathbf{t}-1)}$$

and $\mathbf{q} = q\mathbf{t}^m$, $\mathbf{v}_i = v_i\mathbf{t}^{i-1}$ $(1 \leq i \leq m)$, then $\mathcal{H}_{n,\mathbb{K}}$ is semisimple by Corollary 13.9. Here, the suffix $(\mathbf{t} - 1)$ means localization at this ideal.

Now the restriction rule Theorem 13.6(2) combined with Theorem 13.21(3) gives

$$\text{Res}([S_{\mathbb{K}}^{\lambda}]) = \sum_{\mu:|\lambda/\mu|=1} [S_{\mathbb{K}}^{\mu}].$$

By Frobenius reciprocity and the semisimplicity of $\mathcal{H}_{n,\mathbb{K}}$, this also gives

$$\text{Ind}([S_{\mathbb{K}}^{\lambda}]) = \sum_{\mu:|\mu/\lambda|=1} [S_{\mathbb{K}}^{\mu}].$$

We now apply the decomposition map. We take lattices $S_{\mathbb{S}}^{\lambda}$ and $\bigoplus_{\mu:|\lambda/\mu|=1} S_{\mathbb{S}}^{\mu}$ or lattices $\mathcal{H}_{n+1,\mathbb{S}} \otimes_{\mathcal{H}_{n,\mathbb{S}}} S_{\mathbb{S}}^{\lambda}$ and $\bigoplus_{\mu:|\mu/\lambda|=1} S_{\mathbb{S}}^{\mu}$. Since the decompostion map does not depend on the choice of the $\mathbb{S}$-lattice, tensoring $-\otimes\mathbb{F}$ gives the desired formulas. □

Recall that we have assumed the parameter condition $q = \sqrt[r]{1} \neq 1$, $v_i = q^{\gamma_i}$ $(1 \leq i \leq m)$. With this condition, we may define i-restriction and i-induction as follows.

DEFINITION 13.33. Let M be an $\mathcal{H}_{n,\mathbb{F}}$-module. We denote by $P_{n,\lambda(z)}(M)$ the generalized eigenspace of $c_n(z) = \prod_{k=1}^{n}(z - t_k)$ with respect to the eigenvalue $\lambda(z) = \prod_{k=1}^{n}(z - q^{i_k}) \in \mathbb{F}[z]$. Since $c_n(z)$ commutes with $\mathcal{H}_{n,\mathbb{F}}$, we have

$$P_{n,\lambda(z)}(M) \in \mathcal{H}_{n,\mathbb{F}} - mod.$$

For $i \in \mathbb{Z}/r\mathbb{Z}$, **i-induction** $i-\text{Ind}$ is defined by

$$i-\text{Ind}(M) = \bigoplus_{\lambda(z)\in\mathbb{F}[z]} P_{n+1,\lambda(z)(z-q^i)}\left(\text{Ind}(P_{n,\lambda(z)}(M))\right).$$

Similarly, for $i \in \mathbb{Z}/r\mathbb{Z}$, **i-restriction** $i-\mathrm{Res}$ is defined by

$$i-\mathrm{Res}(M) = \bigoplus_{\lambda(z)\in\mathbb{F}[z]} P_{n-1,\lambda(z)/(z-q^i)}\left(\mathrm{Res}(P_{n,\lambda(z)}(M))\right).$$

LEMMA 13.34. *i-Ind and i-Res are exact functors and they induce the following linear operators.*

$$\begin{array}{rccc} i-\mathrm{Ind}: & K_0(\mathcal{H}_{n,\mathbb{F}}-mod) & \longrightarrow & K_0(\mathcal{H}_{n+1,\mathbb{F}}-mod), \\ & [M] & \mapsto & [i-\mathrm{Ind}(M)] \\ i-\mathrm{Res}: & K_0(\mathcal{H}_{n,\mathbb{F}}-mod) & \longrightarrow & K_0(\mathcal{H}_{n-1,\mathbb{F}}-mod). \\ & [M] & \mapsto & [i-\mathrm{Res}(M)] \end{array}$$

PROOF. By Corollary 13.13, i-Ind and i-Res are compositions of exact functors; hence everything is obvious. □

Recall that we introduced the Fock space $\mathcal{F}_{\gamma_m,\ldots,\gamma_1}$ in Theorem 10.10. We now abuse notation. Let $\mathcal{F}_{\gamma_m,\ldots,\gamma_1}$ denote the specialized space at $v=1$. Thus, $\mathcal{F}_{\gamma_m,\ldots,\gamma_1}$ is now a $\mathbb{Q}$-vector space with basis $\{\lambda=(\lambda^{(m)},\ldots,\lambda^{(1)})\}$. Using the r-residues associated with $(\gamma_m,\ldots,\gamma_1)$, the operators e_i, f_i, h_i, d on $\mathcal{F}_{\gamma_m,\ldots,\gamma_1}$ are defined as follows.

$$e_i\lambda = \sum_{\mu:r(\lambda/\mu)\equiv i} \mu, \quad f_i\lambda = \sum_{\mu:r(\mu/\lambda)\equiv i} \mu, \quad h_i\lambda = N_i(\lambda)\lambda,$$

$$d\lambda = -W_0(\lambda)\lambda.$$

By a specialization argument, or by an explicit verification of the defining relations, we have the following lemma.

LEMMA 13.35. *$\mathcal{F}_{\gamma_m,\ldots,\gamma_1}$ is an integrable $\mathfrak{g}(A^{(1)}_{r-1})$-module. Let $U(\mathfrak{g}(A^{(1)}_{r-1}))\emptyset$ be the $\mathfrak{g}(A^{(1)}_{r-1})$-submodule generated by the empty multi-Young diagram $\emptyset$. Then $U(\mathfrak{g}(A^{(1)}_{r-1}))\emptyset \simeq V(\Lambda)$ where $\Lambda=\sum_{i=1}^m \Lambda_{\gamma_i}$.*

DEFINITION 13.36. Let $\mathcal{H}_{n,\mathbb{F}}$ for $n\ge 0$ be Hecke algebras of type $G(m,1,n)$ with common parameters $q=\sqrt[r]{1}$ $(r\ge 2)$, $v_i=q^{\gamma_i}$ $(1\le i\le m)$. We define $V(\mathbb{F})$ by

$$V_n(\mathbb{F}) = \mathrm{Hom}_{\mathbb{Z}}\left(K_0(\mathcal{H}_{n,\mathbb{F}}-mod),\mathbb{Q}\right),\ V(\mathbb{F}) = \bigoplus_{n\ge 0} V_n(\mathbb{F}).$$

The transpose of the operators i-Ind and i-Res are denoted as follows.

$$e_i := (i-\mathrm{Ind})^T : V(\mathbb{F})\to V(\mathbb{F}),$$
$$f_i := (i-\mathrm{Res})^T : V(\mathbb{F})\to V(\mathbb{F}).$$

LEMMA 13.37. *Let $r(x)$ be the r-residue associated with $(\gamma_m,\ldots,\gamma_1)$, where the γ_i are given in the parameter condition $q=\sqrt[r]{1}$ $(r\ge 2)$, $v_i=q^{\gamma_i}$ $(1\le i\le m)$. Then*

$$i-\mathrm{Ind}([S^\lambda_{\mathbb{F}}]) = \sum_{\mu:r(\mu/\lambda)\equiv i}[S^\mu_{\mathbb{F}}], \quad i-\mathrm{Res}([S^\lambda_{\mathbb{F}}]) = \sum_{\mu:r(\lambda/\mu)\equiv i}[S^\mu_{\mathbb{F}}].$$

PROOF. We prove the formula for i-Ind. The proof for i-Res is similar. Write $P_{\{q^{\mathrm{res}(x)}|x\in\lambda\}}$ for $P_{n,\lambda(z)}$ if $\lambda(z)=\prod_{x\in\lambda}(z-q^{\mathrm{res}(x)})$. Then by Theorem 13.21(2), we have

$$S_{\mathbb{F}}^{\lambda}=P_{\{q^{\mathrm{res}(x)}|x\in\lambda\}}(S_{\mathbb{F}}^{\lambda}).$$

Since Lemma 13.32 implies that

$$\begin{aligned} i-\mathrm{Ind}([S_{\mathbb{F}}^{\lambda}]) &= P_{\{q^{\mathrm{res}(x)}|x\in\lambda\}\cup\{q^i\}}\left(\mathrm{Ind}([S_{\mathbb{F}}^{\lambda}])\right) \\ &= \sum_{\mu:|\mu/\lambda|=1}[P_{\{q^{\mathrm{res}(x)}|x\in\lambda\}\cup\{q^i\}}(S_{\mathbb{F}}^{\mu})], \end{aligned}$$

and $S_{\mathbb{F}}^{\mu}=P_{\{q^{\mathrm{res}(x)}|x\in\mu\}}(S_{\mathbb{F}}^{\mu})$ implies that μ with $\mathrm{res}(\mu/\lambda)\not\equiv i$ all vanish, we have the desired formula. □

PROPOSITION 13.38. *Let $\mathcal{H}_{n,\mathbb{F}}$ for $n\geq 0$ be Hecke algebras of type $G(m,1,n)$ with common parameters $q=\sqrt[r]{1}$ $(r\geq 2)$, $v_i=q^{\gamma_i}$ $(1\leq i\leq m)$ and let $(\mathbb{K},\mathbb{S},\mathbb{F})$ be a modular system with parameters such that $\mathcal{H}_{n,\mathbb{K}}$ is semisimple. Since $\{[S_{\mathbb{K}}^{\lambda}]\}_{\lambda\vdash n}$ is a basis of $K_0(\mathcal{H}_{n,\mathbb{K}}-mod)$, we have its dual basis, which we denote by $\{[S_{\mathbb{K}}^{\lambda}]^*\}_{\lambda\vdash n}$. Then $\bigsqcup_{n\geq 0}\{[S_{\mathbb{K}}^{\lambda}]^*\}_{\lambda\vdash n}$ is a basis of $V(\mathbb{K})$. If we identify $V(\mathbb{K})$ with $\mathcal{F}_{\gamma_m,\dots,\gamma_1}$ via $[S_{\mathbb{K}}^{\lambda}]^*\Leftrightarrow\lambda$, then the following diagram commutes.*

$$\begin{array}{ccccccc} 0 & \longrightarrow & V(\mathbb{F}) & \stackrel{d_{\mathbb{K},\mathbb{F}}^T}{\longrightarrow} & V(\mathbb{K}) & \simeq & \mathcal{F}_{\gamma_m,\dots,\gamma_1} \\ & & e_i,f_i\downarrow & & & & \downarrow e_i,f_i \\ 0 & \longrightarrow & V(\mathbb{F}) & \stackrel{d_{\mathbb{K},\mathbb{F}}^T}{\longrightarrow} & V(\mathbb{K}) & \simeq & \mathcal{F}_{\gamma_m,\dots,\gamma_1} \end{array}$$

Here $d_{\mathbb{K},\mathbb{F}}^T$ is the transpose of the decomposition map $d_{\mathbb{K},\mathbb{F}}$.

PROOF. The injectivity of $d_{\mathbb{K},\mathbb{F}}^T$ follows from Corollary 13.23(2). Take an element $x\in V(\mathbb{F})$ and write

$$d_{\mathbb{K},\mathbb{F}}^T(x)=\sum_{\lambda}c_\lambda[S_{\mathbb{K}}^{\lambda}]^*\quad(c_\lambda\in\mathbb{Q}).$$

Then Corollary 13.23(1) implies that

$$\langle\, d_{\mathbb{K},\mathbb{F}}^T(e_ix),\,[S_{\mathbb{K}}^{\mu}]\,\rangle=\langle\, i-\mathrm{Ind}^T(x),\,[S_{\mathbb{F}}^{\mu}]\,\rangle=\langle\, x,\,i-\mathrm{Ind}([S_{\mathbb{F}}^{\mu}])\,\rangle.$$

We use Lemma 13.37 to obtain

$$\begin{aligned} &=\langle\, x,\sum_{\nu:\mathrm{res}(\nu/\mu)\equiv i}[S_{\mathbb{F}}^{\nu}]\,\rangle=\sum_{\nu:\mathrm{res}(\nu/\mu)\equiv i}\langle\, d_{\mathbb{K},\mathbb{F}}^T(x),\,[S_{\mathbb{K}}^{\nu}]\,\rangle \\ &=\sum_{\nu:\mathrm{res}(\nu/\mu)\equiv i}\langle\sum_{\lambda}c_\lambda[S_{\mathbb{K}}^{\lambda}]^*,\,[S_{\mathbb{K}}^{\nu}]\,\rangle=\sum_{\lambda:\mathrm{res}(\lambda/\mu)\equiv i}c_\lambda. \end{aligned}$$

Hence we have $d_{\mathbb{K},\mathbb{F}}^T(e_ix)=\sum_{\lambda,\mu:\mathrm{res}(\lambda/\mu)\equiv i}c_\lambda[S_{\mathbb{K}}^{\mu}]^*$.

On the other hand, we have

$$e_i\left(\sum_{\lambda}c_\lambda\lambda\right)=\sum_{\lambda}c_\lambda\left(\sum_{\mu:\mathrm{res}(\lambda/\mu)\equiv i}\mu\right)=\sum_{\lambda,\mu:\mathrm{res}(\lambda/\mu)\equiv i}c_\lambda\,\mu.$$

in $\mathcal{F}_{\gamma_m,\dots,\gamma_1}$. Thus the commutativity for e_i follows. The commutativity for f_i may be proved similarly. □

By Proposition 13.38, $V(\mathbb{F})$ is an integrable $\mathfrak{g}'(A^{(1)}_{r-1})$-module. As the rank of the Gram matrix of $S^\mu_\mathbb{F}$ determines whether or not $D^\mu_\mathbb{F}$ is non-zero, the set $\{\mu|\, D^\mu_\mathbb{F}\neq 0\}$ does not change if we replace $\mathbb{F}$ by its extension field. In fact, Theorem 13.21(4),(5) implies that the dimension of $V_n(\mathbb{F})$ is the same. Since the Specht modules span the Grothendieck group $V(\mathbb{F})$ by Corollary 13.23, Lemma 13.37 implies that the module structure of $V(\mathbb{F})$ is determined by the order of q. In particular, if we replace $\mathbb{F}$ by an extension field we have the same $\mathfrak{g}'(A^{(1)}_{r-1})$-module. Hence, without loss of generality, we may assume that $\mathbb{F}$ is algebraically closed in our study of $V(\mathbb{F})$.

We turn to $\mathcal{C}_{n,\mathbb{F}}$, and consider the i-restriction operators for affine Hecke algebras.

DEFINITION 13.39. Let $\hat{H}_{n,\mathbb{F}}$ $(n=0,1,\dots)$ be affine Hecke algebras with a common parameter $q=\sqrt[r]{1}$ $(r\geq 2)$. If $M\in\mathcal{C}_{n,\mathbb{F}}$ then we define $P_{n,\lambda(z)}(M)$ to be the generalized eigenspace of $\hat{c}_n(z)=\prod_{k=1}^n(z-X_k)$ with eigenvalue $\lambda(z)=\prod_{k=1}^n(z-q^{i_k})\in\mathbb{F}[z]$. Since $\hat{c}_n(z)$ commutes with $\hat{H}_{n,\mathbb{F}}$ we have

$$P_{n,\lambda(z)}(M)\in\mathcal{C}_{n,\mathbb{F}}.$$

Abusing notation, for $i\in\mathbb{Z}/r\mathbb{Z}$ we define the i-restriction functor $i-\mathrm{Res}$ for the affine Hecke algebra by

$$i-\mathrm{Res}[M]=\bigoplus_{\lambda(z)\in\mathbb{F}[z]}P_{n-1,\lambda(z)/(z-q^i)}\left(\mathrm{Res}(P_{n,\lambda(z)}(M))\right).$$

This is an exact functor. Similarly, we also denote by $i-\mathrm{Res}$ the induced linear operator

$$i-\mathrm{Res}: K_0(\mathcal{C}_{n,\mathbb{F}})\longrightarrow K_0(\mathcal{C}_{n-1,\mathbb{F}})$$

given by $[M]\mapsto[i-\mathrm{Res}(M)]$. We define $U(\mathbb{F})$ by

$$U_n(\mathbb{F})=\mathrm{Hom}_\mathbb{Z}\left(K_0(\mathcal{C}_{n,\mathbb{F}}),\mathbb{Q}\right),\ U(\mathbb{F})=\bigoplus_{n\geq 0}U_n(\mathbb{F}).$$

The transpose of i-Res is denoted by

$$f_i:=(i-\mathrm{Res})^T: U(\mathbb{F})\to U(\mathbb{F}).$$

LEMMA 13.40. *Let $\mathcal{H}_{n,\mathbb{F}}$ $(n=0,1,\dots)$ be Hecke algebras of type $G(m,1,n)$ with common parameters $q=\sqrt[r]{1}$ $(r\geq 2)$, $v_i=q^{\gamma_i}$ $(1\leq i\leq m)$. Then Lemma 13.27 gives a natural surjective map $U(\mathbb{F})\to V(\mathbb{F})$ with the following properties.*

(1) *The following diagram commutes.*

$$\begin{array}{ccc} U(\mathbb{F}) & \longrightarrow & V(\mathbb{F})\\ f_i\downarrow & & \downarrow f_i\\ U(\mathbb{F}) & \longrightarrow & V(\mathbb{F})\end{array}$$

(2) *Let $\{[D]\}$ be the basis of $\oplus_{n\geq 0}K_0(\mathcal{C}_{n,\mathbb{F}})$ formed by simple modules lying in $\oplus_{n\geq 0}\mathcal{C}_{n,\mathbb{F}}$, and let $\{[D]^*\}$ be its dual basis in $U(\mathbb{F})$. Then*

$$\begin{array}{ccc} U(\mathbb{F}) & \longrightarrow & V(\mathbb{F})\\ [D]^* & \mapsto & \begin{cases}[D]^* & (D\in\mathcal{H}_{n,\mathbb{F}}-\mathit{mod})\\ 0 & (\mathit{otherwise})\end{cases}\end{array}.$$

PROOF. The evaluation of the image of $[D]^*$ on $[M] \in K_0(\mathcal{H}_{n,\mathbb{F}} - mod)$ is $[M : D]$ if we consider M as an $\hat{H}_{n,\mathbb{F}}$-module.

- If D is an $\mathcal{H}_{n,\mathbb{F}}$-module, then we have $[M : D] = \langle [D]^*, [M] \rangle$ by viewing $[D]^*$ as an element of $V(\mathbb{F})$.
- If D is not an $\mathcal{H}_{n,\mathbb{F}}$-module, we have $[M : D] = 0$ since the $\mathcal{H}_{n,\mathbb{F}}$-module M cannot have $[D]$ as a composition factor. In particular, the image of $[D]^*$ is 0.

Hence we have the result. □

Using the preceding arguments we can prove the following proposition. This proposition relates the modular representation theory of the Hecke algebra of type $G(m,1,n)$ to the Hayashi realization of the quantum algebra of type $A^{(1)}_{r-1}$, and plays an important role in the proof of Theorem 12.5 in the next chapter.

PROPOSITION 13.41. *Let $\mathcal{H}_{n,\mathbb{F}}$ ($n \in \mathbb{Z}_{\geq 0}$) be Hecke algebras of type $G(m,1,n)$ with common parameters $q = \sqrt[r]{1}$ ($r \geq 2$), $v_i = q^{\gamma_i}$ ($1 \leq i \leq m$). If $U(\mathbb{F})$ is spanned by the elements which are obtained by applying finitely many $f_0, \ldots, f_{r-1}$ to $U_0(\mathbb{F})$ then we have*

$$\begin{array}{cccccc} d^T_{\mathbb{K},\mathbb{F}}: & V(\mathbb{F}) & \simeq & U(\mathfrak{g}(A^{(1)}_{r-1}))\emptyset & \subset & \mathcal{F}_{\gamma_m,\ldots,\gamma_1} \\ & [\,D^\mu_{\mathbb{F}}\,]^* & \mapsto & \sum_{\lambda \trianglerighteq \mu} d_{\lambda,\mu}\lambda & & \end{array}.$$

In particular, we may identify $V(\mathbb{F})$ with the irreducible $\mathfrak{g}(A^{(1)}_{r-1})$-module $V(\Lambda)$ where $\Lambda = \sum_{i=1}^m \Lambda_{\gamma_i}$.

PROOF. By Lemma 13.40(1) and the assumptions on $U(\mathbb{F})$, the space $V(\mathbb{F})$ is the $\mathfrak{g}(A^{(1)}_{r-1})$-module generated by $V_0(\mathbb{F}) = \mathbb{Q}\emptyset$. Hence we may identify $V(\mathbb{F})$ with $U(\mathfrak{g}(A^{(1)}_{r-1}))\emptyset$. Since $\mathcal{F}_{\gamma_m,\ldots,\gamma_1}$ is integrable, we have $U(\mathfrak{g}(A^{(1)}_{r-1}))\emptyset \simeq V(\Lambda)$.

Finally, the equation

$$\langle\, d^T_{\mathbb{K},\mathbb{F}}([\,D^\mu_{\mathbb{F}}\,]^*),\ [\,S^\lambda_{\mathbb{K}}\,]\rangle = \langle\,[\,D^\mu_{\mathbb{F}}\,]^*,\ [\,S^\lambda_{\mathbb{F}}\,]\rangle = [S^\lambda_{\mathbb{F}} : D^\mu_{\mathbb{F}}] = d_{\lambda,\mu}$$

shows that $d^T_{\mathbb{K},\mathbb{F}}([\,D^\mu_{\mathbb{F}}\,]^*) = \sum_\lambda d_{\lambda,\mu}\,[\,S^\lambda_{\mathbb{K}}\,]^*$. □

To show that the assumptions of Proposition 13.41 actually hold and hence prove the statement in Theorem 12.5 which relates the PIMs to the canonical basis, we use the geometric theory of quantum algebras and affine Hecke algebras developed by Lusztig and Ginzburg. The next chapter begins by explaining this theory and then, after some necessary preparations, gives the promised proof of Theorem 12.5.

CHAPTER 14

The proof of Theorem 12.5

14.1. Representations of a cyclic quiver

The **cyclic quiver** of length r is the directed graph $\Gamma = (\mathcal{V}, \mathcal{E})$ with vertices $\mathcal{V} = \mathbb{Z}/r\mathbb{Z}$ and directed edges $\mathcal{E}$ given by

$$\mathcal{E} = \{(i \pmod r), i+1 \pmod r)) | \, i \in \mathbb{Z}/r\mathbb{Z}\} \subset \mathcal{V} \times \mathcal{V}.$$

We assume that $r \geq 2$. The length r of the quiver will be identified with the r in the suffix of $\mathfrak{g}(A^{(1)}_{r-1})$.

Let $p > 0$ be a prime which does not divide r, $\mathbb{F}_{p^e}$ be the finite field with p^e elements. We fix another prime $l \neq p$ and consider the algebraic closure $\overline{\mathbb{Q}}_l$ of the local field $\mathbb{Q}_l$.

DEFINITION 14.1. A **representation** V of the quiver Γ is a collection of $\overline{\mathbb{F}}_p$-vector spaces and $\overline{\mathbb{F}}_p$-linear maps

$$V = (\{V_i\}_{i \in \mathbb{Z}/r\mathbb{Z}}, \{x_{i,i+1}\}_{i \in \mathbb{Z}/r\mathbb{Z}})$$

where $x_{i,i+1} \in \operatorname{Hom}_{\overline{\mathbb{F}}_p}(V_i, V_{i+1})$.

The **dimension vector** of V is the vector $\mathbf{d} = (d_0, \ldots, d_{r-1}) \in \mathbb{Z}^r_{\geq 0}$ where $d_i = \dim_{\overline{\mathbb{F}}_p} V_i$ for $0 \leq i < r$.

DEFINITION 14.2. Let V and W be representations of Γ given by

$$V = (\{V_i\}_{i \in \mathbb{Z}/r\mathbb{Z}}, \{x_{i,i+1}\}_{i \in \mathbb{Z}/r\mathbb{Z}}), \quad W = (\{W_i\}_{i \in \mathbb{Z}/r\mathbb{Z}}, \{y_{i,i+1}\}_{i \in \mathbb{Z}/r\mathbb{Z}}).$$

A **homomorphism** from V to W is a set $\Phi = \{\phi_i\}_{i \in \mathbb{Z}/r\mathbb{Z}}$ of linear maps where

$$\phi_i \in \operatorname{Hom}_{\overline{\mathbb{F}}_p}(V_i, W_i) \ \ (i \in \mathbb{Z}/r\mathbb{Z})$$

and such that $\phi_{i+1} \circ x_{i,i+1} = y_{i,i+1} \circ \phi_i$ for all $i \in \mathbb{Z}/r\mathbb{Z}$.

If every ϕ_i, for $i \in \mathbb{Z}/r\mathbb{Z}$, is a linear isomorphism then we say that Φ is an **isomorphism**. If there exists an isomorphism between two representations we say that these representations are **isomorphic**.

DEFINITION 14.3. Let V and W be representations of Γ as above. The **direct sum** of these representations is the representation

$$V \oplus W = \left(\{V_i \oplus W_i\}_{i \in \mathbb{Z}/r\mathbb{Z}}, \{x_{i,i+1} \oplus y_{i,i+1}\}_{i \in \mathbb{Z}/r\mathbb{Z}}\right).$$

DEFINITION 14.4. Let V and W be representations of Γ. If $V_i \subset W_i$ and $y_{i,i+1}|_{V_i} = x_{i,i+1}$ for all $i \in \mathbb{Z}/r\mathbb{Z}$ then V is a **subrepresentation** of W.

Assume that V is a subrepresentation of W and let $\bar{y}_{i,i+1}$ be the linear map $W_i/V_i \to W_{i+1}/V_{i+1}$ induced by $y_{i,i+1}$. Then the representation

$$W/V = (\{W_i/V_i\}_{i \in \mathbb{Z}/r\mathbb{Z}}, \{\bar{y}_{i,i+1}\}_{i \in \mathbb{Z}/r\mathbb{Z}})$$

is the **quotient representation** of W by V.

Instead of specifying the set of vector spaces $\{V_i\}_{i\in\mathbb{Z}/r\mathbb{Z}}$ we often consider their direct sum

$$V = \bigoplus_{i\in\mathbb{Z}/r\mathbb{Z}} V_i,$$

which we view as a $\mathbb{Z}/r\mathbb{Z}$-graded vector space.

The set of linear maps $\{x_{i,i+1}\}_{i\in\mathbb{Z}/r\mathbb{Z}}$ is replaced by

$$x = \bigoplus_{i\in\mathbb{Z}/r\mathbb{Z}} x_{i,i+1},$$

which we view as a degree 1 linear endomorphism of the $\mathbb{Z}/r\mathbb{Z}$-graded vector space V. Thus a representation of Γ is a pair (V, x) consisting of a $\mathbb{Z}/r\mathbb{Z}$-graded vector space and a degree 1 linear endomorphism of it.

If x is nilpotent the representation is called **nilpotent**. The **trivial** representation of the quiver is the representation with dimension vector $\mathbf{d} = \mathbf{0}$.

An **indecomposable** representation is a representation which is not isomorphic to a direct sum of two non-trivial representations. A representation is **irreducible** if no non-trivial proper subrepresentations exist.

If we replace $\overline{\mathbb{F}}_p$ by $\mathbb{F}_{p^e}$ then we obtain **representations of Γ over $\mathbb{F}_{p^e}$**. All of the definitions above also make sense for the representations over $\mathbb{F}_{p^e}$.

DEFINITION 14.5. Let $\mathbf{d} \in \mathbb{Z}^r_{\geq 0}$. $V_{\mathbf{d}}$ is the $\mathbb{Z}/r\mathbb{Z}$-graded vector space

$$V = \bigoplus_{i\in\mathbb{Z}/r\mathbb{Z}} V_i$$

such that $V_i = \overline{\mathbb{F}}_p^{d_i}$ for all i. Similarly, we define $V_{\mathbf{d}}^{(e)}$ by

$$V_{\mathbf{d}}^{(e)} = \bigoplus_{i\in\mathbb{Z}/r\mathbb{Z}} V_i^{(e)}$$

where $V_i^{(e)} = \mathbb{F}_{p^e}^{d_i}$. This is a $\mathbb{Z}/r\mathbb{Z}$-graded vector space over $\mathbb{F}_{p^e}$.

Note that the dimension vector of $V_{\mathbf{d}}$ is $\mathbf{d}$. Assume that $V = V_{\mathbf{d}}$. Then the vector space

$$\bigoplus_{i\in\mathbb{Z}/r\mathbb{Z}} \mathrm{Hom}_{\overline{\mathbb{F}}_p}(V_i, V_{i+1}) = \bigoplus_{i\in\mathbb{Z}/r\mathbb{Z}} \mathrm{Hom}_{\overline{\mathbb{F}}_p}(\overline{\mathbb{F}}_p^{d_i}, \overline{\mathbb{F}}_p^{d_{i+1}})$$

is the space of block diagonal matrices $\oplus_{i\in\mathbb{Z}/r\mathbb{Z}}\mathrm{Mat}(d_i, d_{i+1}, \overline{\mathbb{F}}_p)$, so the space is equipped with a natural $\mathbb{F}_{p^e}$-structure: the Frobenius map acts on

$$x \in \bigoplus_{i\in\mathbb{Z}/r\mathbb{Z}} \mathrm{Hom}_{\overline{\mathbb{F}}_p}(V_i, V_{i+1})$$

by raising all entries of x to the p^eth power. Thus the set of the $\mathbb{F}_{p^e}$-rational points of the vector space is equal to

$$\bigoplus_{i\in\mathbb{Z}/r\mathbb{Z}} \mathrm{Mat}(d_i, d_{i+1}, \mathbb{F}_{p^e}) = \bigoplus_{i\in\mathbb{Z}/r\mathbb{Z}} \mathrm{Hom}_{\mathbb{F}_{p^e}}(\mathbb{F}_{p^e}^{d_i}, \mathbb{F}_{p^e}^{d_{i+1}}).$$

In other words, the set of representations of Γ over $\mathbb{F}_{p^e}$ of the form $(V_{\mathbf{d}}^{(e)}, x)$ is the same as the $\mathbb{F}_{p^e}$-rational points of $\oplus_{i\in\mathbb{Z}/r\mathbb{Z}}\mathrm{Hom}_{\overline{\mathbb{F}}_p}(V_i, V_{i+1})$ where $V = V_{\mathbf{d}}$.

We will prove in Lemma 14.8 that the complete set of isomorphism classes of nilpotent representations of Γ does not change when we extend scalars. Thus, any nilpotent representation over $\overline{\mathbb{F}}_p$ (resp. $\mathbb{F}_{p^e}$) is isomorphic to the representation obtained from a representation over $\mathbb{F}_p$ by extension of scalars to $\overline{\mathbb{F}}_p$ (resp. $\mathbb{F}_{p^e}$).

DEFINITION 14.6. For each dimension vector $\mathbf{d}$ we set

$$V_{\mathbf{d}} = \bigoplus_{i\in\mathbb{Z}/r\mathbb{Z}} V_i \quad (V_i = \overline{\mathbb{F}}_p^{d_i})$$

as above. $\mathcal{N}_{V_{\mathbf{d}}}$ is the set of nilpotent degree 1 endomorphisms $x = \oplus_{i\in\mathbb{Z}/r\mathbb{Z}} x_{i,i+1}$ of $V_{\mathbf{d}}$. This is a closed subvariety of the linear variety $\mathrm{End}_{\overline{\mathbb{F}}_p}(V_{\mathbf{d}})$.

The algebraic group $G_{V_{\mathbf{d}}} = \prod_{i\in\mathbb{Z}/r\mathbb{Z}} GL(d_i, \overline{\mathbb{F}}_p)$ acts on $\mathcal{N}_{V_{\mathbf{d}}}$ by conjugation.

Note that each of $\mathcal{N}_{V_{\mathbf{d}}}$ and $G_{V_{\mathbf{d}}}$ is equipped with a natural $\mathbb{F}_{p^e}$-structure.

It is clear that $G_{V_{\mathbf{d}}}$-orbits of $\mathcal{N}_{V_{\mathbf{d}}}$ are in bijection with isomorphism classes of nilpotent representations with dimension vector $\mathbf{d}$. It is not hard to describe the $G_{V_{\mathbf{d}}}$-orbits of $\mathcal{N}_{V_{\mathbf{d}}}$ explicitly. To do this, we introduce the notion of multisegments.

DEFINITION 14.7. A **segment** of length l is a sequence of l consecutive residues (taking values in $\mathbb{Z}/r\mathbb{Z}$)

x_1	x_2	$\cdots$	x_l

where $x_i = k+i-1$ $(1 \le i \le l)$ for some $k \in \mathbb{Z}/r\mathbb{Z}$. We call the residue x_l the **residue** of the segment.

To each segment we associate an indecomposable representation (V, x) of Γ by defining

$$V_i = \bigoplus_{j:k+j\equiv i \pmod r} \overline{\mathbb{F}}_p u_j, \quad x(u_j) = \begin{cases} u_{j+1} & (j+1 < l) \\ 0 & (otherwise) \end{cases},$$

where $V = \oplus_{i\in\mathbb{Z}/r\mathbb{Z}} V_i = \oplus_{0\le j<l} \overline{\mathbb{F}}_p u_j$ and we identify each component V_i with $\overline{\mathbb{F}}_p^{d_i}$ $(d_i = \dim_{\overline{\mathbb{F}}_p} V_i)$ using the basis $\{u_j\}_{j:k+j\equiv i \pmod r}$.

A **multisegment** is a multiset of segments. The **size** of a multisegment $\underline{m}$ is the sum of the lengths of segments in $\underline{m}$.

Since each segment in the multisegment is associated with an indecomposable representation of Γ, we associate their direct sum to the multisegment. If $\underline{m}$ is a multisegment, then we denote the corresponding representation of Γ by $V_{\underline{m}} = (V_{\mathbf{d}}, x_{\underline{m}})$, and denote the associated $G_{V_{\mathbf{d}}}$-orbit $G_{V_{\mathbf{d}}} x_{\underline{m}}$ by $\mathcal{O}_{\underline{m}}$. The dimension vector $\mathbf{d}$ is also called the **dimension vector** of $\underline{m}$.

Note that the linear map $x_{\underline{m}}$ is a $\mathbb{F}_p$-rational point of $\mathcal{N}_{V_{\mathbf{d}}}$.

LEMMA 14.8. *The $V_{\underline{m}}$ form a complete set of isomorphism classes of nilpotent representations. In particular, we have*

$$\mathcal{N}_{V_{\mathbf{d}}} = \bigsqcup_{\underline{m}} \mathcal{O}_{\underline{m}}$$

where $\underline{m}$ runs through all multisegments with dimension vector $\mathbf{d}$.

PROOF. Let (V, x) be a nilpotent representation. Define $s \in \mathrm{End}(V)$ by $s|_{V_i} = \eta^i$ where $\eta = \sqrt[r]{1} \in \overline{\mathbb{F}}_p$. Then $sxs^{-1} = \eta x$. In particular, $\mathrm{Ker}(x^k)$ is stable under s, and

$$\mathrm{Ker}(x^k) = \bigoplus_{i \in \mathbb{Z}/r\mathbb{Z}} \mathrm{Ker}(x^k) \cap V_i.$$

Note that this is also valid for representations over $\mathbb{F}_{p^e}$ if we take $\mathbb{F}_{p^e}$-rational points of $\mathrm{Ker}(x^k)$.

Let N be an integer such that

$$0 \subset \mathrm{Ker}(x) \subset \cdots \subset \mathrm{Ker}(x^{N-1}) \subset \mathrm{Ker}(x^N) = V.$$

We choose eigenvectors $\{v_1, \ldots, v_{j_1}\}$ of s which form the pullback of a basis of $V/\mathrm{Ker}(x^{N-1})$. This is possible since this is the same as pulling back a basis of $V_i/(\mathrm{Ker}(x^{N-1}) \cap V_i)$ to V_i for each i. The vectors

$$\{v_1, \ldots, v_{j_1}; xv_1, \ldots, xv_{j_1}; \ldots; x^{N-1}v_1, \ldots, x^{N-1}v_{j_1}\} \tag{14.1}$$

are linearly independent. To see this it is enough to show that

$$\{x^{N-1}v_1, \ldots, x^{N-1}v_{j_1}\}$$

is linearly independent; however, $\sum_{i=1}^{j_1} c_i x^{N-1} v_i = 0$ implies that $\sum_{i=1}^{j_1} c_i v_i \in \mathrm{Ker}(x^{N-1})$ and thus all $c_i = 0$ by the choice of $v_1, \ldots, v_{j_1}$.

Hence the set (14.1) determines a subrepresentation of (V, x) which is isomorphic to the direct sum of indecomposable representations associated with segments of length N. Starting with these vectors we now construct a basis of V by the following inductive procedure.

Choose vectors $\{v_{j_{k-1}+1}, \ldots, v_{j_k}\}$ from the eigenvectors of s in $\mathrm{Ker}(x^{N-k+1})$ so that they form the pullback of a basis of

$$\mathrm{Ker}(x^{N-k+1}) \Big/ \left(\mathrm{Ker}(x^{N-k}) + \sum_{j>0} x^j \left(\overline{\mathbb{F}}_p v_{j_{k-j-1}+1} + \cdots + \overline{\mathbb{F}}_p v_{j_{k-j}} \right) \right).$$

We consider the vectors

$$\{v_{j_{k-1}+1}, \ldots, v_{j_k}; xv_{j_{k-1}+1}, \ldots, xv_{j_k}; \ldots; x^{N-k}v_{j_{k-1}+1}, \ldots, x^{N-k}v_{j_k}\}.$$

Then the reasoning above shows that these vectors are linearly independent and that they give a subrepresentation isomorphic to the direct sum of the indecomposable representations associated with the segments of length $N-k+1$.

This construction implies that any nilpotent representation of Γ is isomorphic to $V_{\underline{m}}$ for some $\underline{m}$.

Next we prove that different multisegments give non-isomorphic representations. Assume that two representations $V_{\underline{m}}$ and $V_{\underline{m}'}$ are isomorphic. Observe that the dimension of the vector space $(\mathrm{Ker}(x^k) \cap V_i)/(\mathrm{Ker}(x^k) \cap \mathrm{Im}(x) \cap V_i)$ is equal to the number of segments which start with i and have length less than or equal to k. Hence $\underline{m} = \underline{m}'$. □

We note that the same proof works if we replace $\overline{\mathbb{F}}_p$ with $\mathbb{F}_{p^e}$.

DEFINITION 14.9. The $\mathbb{F}_{p^e}$-rational points of $G_{V_{\mathbf{d}}}$, $\mathcal{N}_{V_{\mathbf{d}}}$ and $\mathcal{O}_{\underline{m}}$ are $G_{V_{\mathbf{d}}}(\mathbb{F}_{p^e})$, $\mathcal{N}_{V_{\mathbf{d}}}(\mathbb{F}_{p^e})$ and $\mathcal{O}_{\underline{m}}(\mathbb{F}_{p^e})$ respectively.

COROLLARY 14.10. *Let*

$$\mathcal{N}_{V_{\mathbf{d}}}(\mathbb{F}_{p^e}) = \bigsqcup_{\underline{m}} \mathcal{O}_{\underline{m}}(\mathbb{F}_{p^e}) \subset \bigoplus_{i \in \mathbb{Z}/r\mathbb{Z}} \mathrm{Hom}_{\mathbb{F}_{p^e}}(\mathbb{F}_{p^e}^{d_i}, \mathbb{F}_{p^e}^{d_{i+1}})$$

be the decomposition obtained from Lemma 14.8 by taking $\mathbb{F}_{p^e}$-rational points. Then each $\mathcal{O}_{\underline{m}}(\mathbb{F}_{p^e})$ is a $G_{V_{\mathbf{d}}}(\mathbb{F}_{p^e})$-orbit.

14.2. The Hall algebra and the quantum algebra

We review the construction of the Hall algebra following [**cb-E**].

DEFINITION 14.11. Let Z_1, Z_2 be finite sets and $\pi : Z_1 \to Z_2$ a map. Given a function f on Z_1, the **pushforward** of f is the function $\pi_! f$ on Z_2 defined by

$$(\pi_! f)(z_2) = \sum_{z_1 \in \pi^{-1}(z_2)} f(z_1).$$

If a function f on Z_2 is given, then the **pullback** $\pi^* f$ of f is the function on Z_1 defined by

$$(\pi^* f)(z_1) = f(\pi(z_1)).$$

DEFINITION 14.12. For each $\mathbf{d} \in \mathbb{Z}_{\geq 0}^r$ let $\mathcal{H}_{V_{\mathbf{d}}}(\mathbb{F}_{p^e})$ be the $\overline{\mathbb{Q}}_l$-vector space of $G_{V_{\mathbf{d}}}(\mathbb{F}_{p^e})$-invariant $\overline{\mathbb{Q}}_l$-valued functions on $\mathrm{End}_{\overline{\mathbb{F}}_p}(V_{\mathbf{d}})$ which have supports in

$$\mathcal{N}_{V_{\mathbf{d}}}(\mathbb{F}_{p^e}) = \bigsqcup_{\underline{m} : \mathcal{O}_{\underline{m}} \subset \mathcal{N}_{V_{\mathbf{d}}}} \mathcal{O}_{\underline{m}}(\mathbb{F}_{p^e}).$$

The characteristic function of $\mathcal{O}_{\underline{m}}(\mathbb{F}_{p^e})$ is

$$1_{\mathcal{O}_{\underline{m}}(\mathbb{F}_{p^e})}(x) = \begin{cases} 1 & (x \in \mathcal{O}_{\underline{m}}(\mathbb{F}_{p^e})) \\ 0 & (otherwise) \end{cases}.$$

The set $\{1_{\mathcal{O}_{\underline{m}}(\mathbb{F}_{p^e})}\}_{\underline{m} : \mathcal{O}_{\underline{m}} \subset \mathcal{N}_{V_{\mathbf{d}}}}$ is a basis of $\mathcal{H}_{V_{\mathbf{d}}}(\mathbb{F}_{p^e})$. Let $\mathbf{t}, \mathbf{w}$ be dimension vectors such that $\mathbf{d} = \mathbf{t} + \mathbf{w}$. We define T, W, V by

$$T = V_{\mathbf{t}}, \quad W = V_{\mathbf{w}}, \quad V = V_{\mathbf{d}},$$

and consider the following diagram.

$$\begin{array}{ccccccc} \mathcal{N}_T \times \mathcal{N}_W & \xleftarrow{p_1} & E'_{T,W} & \xrightarrow{p_2} & E''_{T,W} & \xrightarrow{p_3} & \mathcal{N}_V \\ (x_T, x_W) & \leftarrow & (x, \varphi, \psi) & \mapsto & (x, \varphi(W)) & \mapsto & x \end{array} \tag{14.2}$$

In the diagram, $E'_{T,W} = \{(x, \varphi, \psi)\}$ is the set of triples where $x \in \mathcal{N}_V$ and φ and ψ are linear maps which preserve the $\mathbb{Z}/r\mathbb{Z}$-grading and such that

$$0 \to W \xrightarrow{\varphi} V \xrightarrow{\psi} T \to 0 \text{ (exact)},$$

$$x\varphi(W) \subset \varphi(W),$$

and $E''_{T,W} = \{(x, U)\}$ is the set of pairs where $x \in \mathcal{N}_V$ and $U = \oplus_{i \in \mathbb{Z}/r\mathbb{Z}} U_i$ is a $\mathbb{Z}/r\mathbb{Z}$-graded subspace of V such that

$$\dim U_i = w_i, \quad xU \subset U.$$

The elements x_T, x_W are given by

$$\begin{array}{lllllll} x_T : & T \overset{\bar{\psi}^{-1}}{\simeq} & V/\varphi(W) & \xrightarrow{\bar{x}} & V/\varphi(W) & \overset{\bar{\psi}}{\simeq} & T, \\ x_W : & W \overset{\varphi}{\simeq} & \varphi(W) & \xrightarrow{x} & \varphi(W) & \overset{\varphi^{-1}}{\simeq} & W, \end{array}$$

where $\bar{\psi} : V/\varphi(W) \to T$ and $\bar{x} : V/\varphi(W) \to V/\varphi(W)$ are the linear maps induced by ψ and x respectively.

Note that both $E'_{T,W}$ and $E''_{T,W}$ are equipped with natural $\mathbb{F}_{p^e}$-structures. Lang's theorem applied to $GL(w_i, \overline{\mathbb{F}}_p)$ for all i implies that if U is stable under the Frobenius map then U is obtained from a representation over $\mathbb{F}_{p^e}$ by extension of scalars. Such a representation over $\mathbb{F}_{p^e}$ is uniquely determined. Hence if we replace $\overline{\mathbb{F}}_p$ with $\mathbb{F}_{p^e}$ in the definition of $E'_{T,W}$ (resp. $E''_{T,W}$) above, we obtain the $\mathbb{F}_{p^e}$-rational points $E'_{T,W}(\mathbb{F}_{p^e})$ of $E'_{T,W}$ (resp. $E''_{T,W}(\mathbb{F}_{p^e})$ of $E''_{T,W}$).

In this section, we consider $\mathbb{F}_{p^e}$-rational points of the sets in the diagram (14.2). Thus we have maps between finite sets.

$$\mathcal{N}_T(\mathbb{F}_{p^e}) \times \mathcal{N}_W(\mathbb{F}_{p^e}) \xleftarrow{p_1} E'_{T,W}(\mathbb{F}_{p^e}) \xrightarrow{p_2} E''_{T,W}(\mathbb{F}_{p^e}) \xrightarrow{p_3} \mathcal{N}_V(\mathbb{F}_{p^e})$$

Hence we may define $p_1^*, p_2^*, p_{3!}$, and we have:

LEMMA 14.13. *Let T, W, V and p_1, p_2 be as above. Given two functions $f \in \mathcal{H}_T(\mathbb{F}_{p^e})$ and $g \in \mathcal{H}_W(\mathbb{F}_{p^e})$, define f'' by*

$$f'' = \frac{1}{|G_T(\mathbb{F}_{p^e})||G_W(\mathbb{F}_{p^e})|} p_{2!} p_1^*(f \otimes g).$$

Then $p_1^*(f \otimes g) = p_2^* f''$.

PROOF. First we compute $f''(x, U)$. By definition, $f''(x, U)$ is equal to

$$\frac{1}{|G_T(\mathbb{F}_{p^e})||G_W(\mathbb{F}_{p^e})|} \sum_{(x', \varphi', \psi') \in p_2^{-1}(x,U)} p_1^*(f \otimes g)(x', \varphi', \psi').$$

Fix $\varphi_o : W \simeq U$ and $\psi_o : V \to V/U \simeq T$. Then we have $(x, \varphi_o, \psi_o) \in p_2^{-1}(x, U)$ and

$$(x', \varphi', \psi') \in p_2^{-1}(x, U) \Longleftrightarrow x' = x,\ \varphi'(W) = U = \varphi_o(W)$$

$$\Longleftrightarrow \begin{cases} x' = x & \\ \varphi' = \varphi_o \circ h & (\text{for some } h \in G_W(\mathbb{F}_{p^e})) \\ \psi' = h' \circ \psi_o & (\text{for some } h' \in G_T(\mathbb{F}_{p^e})) \end{cases}.$$

Thus $f''(x, U)$ is equal to

$$\frac{1}{|G_T(\mathbb{F}_{p^e})||G_W(\mathbb{F}_{p^e})|} \sum_{h \in G_W(\mathbb{F}_{p^e})} \sum_{h' \in G_T(\mathbb{F}_{p^e})} p_1^*(f \otimes g)(x, \varphi_o \circ h, h' \circ \psi_o).$$

We consider

$$\begin{array}{lllllll} x'_T : & T \overset{h' \circ \psi_o}{\longleftarrow} & V/U & \xrightarrow{x} & V/U & \overset{h' \circ \psi_o}{\longrightarrow} & T, \\ x'_W : & W \overset{\varphi_o \circ h}{\longrightarrow} & U & \xrightarrow{x} & U & \overset{\varphi_o \circ h}{\longleftarrow} & W. \end{array}$$

Denote x'_T, x'_W by x^o_T, x^o_W if h and h' are the identities of $G_W(\mathbb{F}_{p^e})$ and $G_T(\mathbb{F}_{p^e})$ respectively. Then

$$x'_T = h' x^o_T h'^{-1}, \quad x_W = h^{-1} x^o_W h.$$

As f and g are $G_T(\mathbb{F}_{p^e})$- and $G_W(\mathbb{F}_{p^e})$-invariant functions respectively, we have

$$p_1^*(f \otimes g)(x, \varphi_o \circ h, h' \circ \psi_o) = f(x'_T)g(x'_W) = f(x^o_T)g(x^o_W).$$

These values are independent of h and h'. Hence we have the following formula:

$$f''(x, U) = f(x^o_T)g(x^o_W). \tag{14.3}$$

Now we prove that $p_1^*(f \otimes g) = p_2^* f''$. Apply the formula (14.3) to

$$p_2^* f''(x, \varphi, \psi) = f''(x, \varphi(W)).$$

As $U = \varphi(W)$, we may assume that $\varphi_o = \varphi, \psi_o = \psi$ and $x^o_T = x_T, x^o_W = x_W$. Hence we have

$$f''(x, \varphi(W)) = f(x_T)g(x_W) = p_1^*(f \otimes g)(x, \varphi, \psi),$$

proving the result. □

LEMMA 14.14. *Let T, W, V and p_1, p_2, p_3 be as above and given two functions $f \in \mathcal{H}_T(\mathbb{F}_{p^e})$ and $g \in \mathcal{H}_W(\mathbb{F}_{p^e})$ define f'' as in Lemma 14.13. Then*

$$p_{3!} f'' \in \mathcal{H}_V(\mathbb{F}_{p^e}).$$

PROOF. Since $(x', U') \in p_3^{-1}(x) \Leftrightarrow x' = x$ and $xU' \subset U'$, we have

$$p_{3!} f''(x) = \sum_{U : xU \subset U} f''(x, U).$$

On the other hand, we have

$$(x', U') \in p_3^{-1}(hxh^{-1}) \Leftrightarrow x' = hxh^{-1},\ U' = hU,\ xU \subset U$$

for hxh^{-1} ($h \in G_V(\mathbb{F}_{p^e})$). This implies that

$$p_{3!} f''(hxh^{-1}) = \sum_{U : xU \subset U} f''(hxh^{-1}, hU).$$

Therefore, it is enough to prove that $f''(hxh^{-1}, hU) = f''(x, U)$.

As in the proof of Lemma 14.13, we take $(x, \varphi_o, \psi_o) \in p_2^{-1}(x, U)$ and fix x^o_T, x^o_W. Then we have $(hxh^{-1}, h \circ \varphi_o, \psi_o \circ h^{-1}) \in p_2^{-1}(hxh^{-1}, hU)$. The map p_1 sends this triple to (x^o_T, x^o_W) because

$$\begin{array}{llllllll} x^o_T: & T & \overset{\psi_o \circ h^{-1}}{\longleftarrow} & V/hU & \overset{hxh^{-1}}{\longrightarrow} & V/hU & \overset{\psi_o \circ h^{-1}}{\longrightarrow} & T, \\ x^o_W: & W & \overset{h \circ \varphi_o}{\longrightarrow} & hU & \overset{hxh^{-1}}{\longrightarrow} & hU & \overset{h \circ \varphi_o}{\longleftarrow} & W. \end{array}$$

Thus we get $f''(hxh^{-1}, hU) = f(x^o_T)g(x^o_W) = f''(x, U)$ by the formula (14.3), which proves that $p_{3!} f''$ is $G_{V_{\mathbf{d}}}(\mathbb{F}_{p^e})$-invariant. □

DEFINITION 14.15. Let T, W, V and p_1, p_2, p_3 be as above and given two functions $f \in \mathcal{H}_T(\mathbb{F}_{p^e})$ and $g \in \mathcal{H}_W(\mathbb{F}_{p^e})$ define f'' as in Lemma 14.13. Then the product fg is defined by

$$fg = p^{-e(\sum t_i w_{i+1} + \sum t_i w_i)/2} p_{3!} f'' \in \mathcal{H}_V(\mathbb{F}_{p^e}).$$

Recall that we have been working with representations of Γ over $\mathbb{F}_{p^e}$ in this section. Thus we may define the following integers $F^{\underline{n}}_{\underline{m}, \underline{m}'}(\mathbb{F}_{p^e})$.

DEFINITION 14.16. We write $U \subset V_{\underline{n}}$ if U is a subrepresentation of $V_{\underline{n}}$, i.e. if U is a $\mathbb{Z}/r\mathbb{Z}$-graded $\mathbb{F}_{p^e}$-subspace of $V_{\underline{n}}$ and $x_{\underline{n}}U \subset U$. Then $F^{\underline{n}}_{\underline{m},\underline{m}'}(\mathbb{F}_{p^e})$ is the cardinality of a finite set defined by

$$F^{\underline{n}}_{\underline{m},\underline{m}'}(\mathbb{F}_{p^e}) = \left|\{U \subset V_{\underline{n}} | U \simeq V_{\underline{m}'},\ V_{\underline{n}}/U \simeq V_{\underline{m}}\}\right|.$$

In other words, $F^{\underline{n}}_{\underline{m},\underline{m}'}(\mathbb{F}_{p^e})$ is the number of subrepresentations U of $V_{\underline{n}}$ which are isomorphic to $V_{\underline{m}'}$ such that $V_{\underline{n}}/U$ is isomorphic to $V_{\underline{m}}$.

DEFINITION 14.17. The algebra $\mathcal{H}(\mathbb{F}_{p^e})$ is the vector space

$$\mathcal{H}(\mathbb{F}_{p^e}) = \bigoplus_{\mathbf{d}\in\mathbb{Z}^r_{\geq 0}} \mathcal{H}_{V_{\mathbf{d}}}(\mathbb{F}_{p^e})$$

equipped with the product defined in Definition 14.15. We set $u_{\underline{m}} = 1_{\mathcal{O}_{\underline{m}}(\mathbb{F}_{p^e})}$.

The $u_{\underline{m}}$ form a basis of $\mathcal{H}(\mathbb{F}_{p^e})$.

PROPOSITION 14.18. *$\mathcal{H}(\mathbb{F}_{p^e})$ is a unital associative algebra.*

PROOF. Since the $u_{\underline{m}}$ form a basis of $\mathcal{H}(\mathbb{F}_{p^e})$, it is enough to consider products of these basis elements in order to prove associativity.

Write

$$u_{\underline{m}}u_{\underline{m}'} = \sum c^{\underline{n}}_{\underline{m},\underline{m}'}u_{\underline{n}}.$$

To evaluate the coefficients, we take an element $x_{\underline{n}} \in \mathcal{N}_{V_{\mathbf{d}}}$ of $\mathcal{O}_{\underline{n}}(\mathbb{F}_{p^e})$, where $\mathbf{d}$ is the dimension vector of $\underline{n}$. Then

$$c^{\underline{n}}_{\underline{m},\underline{m}'} = u_{\underline{m}}u_{\underline{m}'}(x_{\underline{n}}).$$

To compute the right hand side, we apply formula (14.3) to $f = u_{\underline{m}}$ and $g = u_{\underline{m}'}$. For a subrepresentation U of $V_{\underline{n}}$ we have

$$f''(x_{\underline{n}}, U) = \begin{cases} 1 & (U \simeq V_{\underline{m}'}, V_{\underline{n}}/U \simeq V_{\underline{m}}) \\ 0 & (otherwise) \end{cases}.$$

Since $p_{3!}f''(x_{\underline{n}}) = \sum_{U:x_{\underline{n}}U\subset U} f''(x_{\underline{n}}, U)$, we have

$$c^{\underline{n}}_{\underline{m},\underline{m}'} = p^{-e(\sum t_iw_{i+1}+\sum t_iw_i)/2}F^{\underline{n}}_{\underline{m},\underline{m}'}(\mathbb{F}_{p^e}).$$

Let $\mathbf{t}, \mathbf{w}, \mathbf{v}$ be the dimension vectors of $\underline{m}, \underline{m}', \underline{m}''$. If we compute $(u_{\underline{m}}u_{\underline{m}'})u_{\underline{m}''}$ and $u_{\underline{m}}(u_{\underline{m}'}u_{\underline{m}''})$ then the coefficient of $u_{\underline{n}}$ is the same in both cases, and is equal to

$$p^{-e(\sum t_iw_{i+1}+t_iv_{i+1}+w_iv_{i+1}+\sum t_iw_i+t_iv_i+w_iv_i)/2}F^{\underline{n}}_{\underline{m},\underline{m}',\underline{m}''}(\mathbb{F}_{p^e}),$$

where $F^{\underline{n}}_{\underline{m},\underline{m}',\underline{m}''}(\mathbb{F}_{p^e})$ is the number of pairs (U_1, U_2) of subrepresentations of $V_{\underline{n}}$ such that $0 \subset U_1 \subset U_2 \subset V_{\underline{n}}$ and

$$U_1 \simeq V_{\underline{m}''},\ U_2/U_1 \simeq V_{\underline{m}'},\ V_{\underline{n}}/U_2 \simeq V_{\underline{m}}.$$

Hence, the multiplication is associative. □

DEFINITION 14.19. For each $i \in \mathbb{Z}/r\mathbb{Z}$, we denote by S_i the $\mathbb{Z}/r\mathbb{Z}$-graded vector space with dimension vector $(d_j)_{0\leq j\leq r-1} = (\delta_{ij})_{0\leq j\leq r-1}$. We have $\mathcal{N}_{S_i} = \{0\}$. So there exists a unique (irreducible) representation $(S_i, 0)$ of Γ with this dimension vector, which we also denote by S_i. We define $f_i \in \mathcal{H}(\mathbb{F}_{p^e})$ by

$$f_i = 1_{\mathcal{N}_{S_i}}.$$

The following lemma is due to Ringel.

LEMMA 14.20. *The f_i ($i \in \mathbb{Z}/r\mathbb{Z}$) satisfy the Serre relations. Namely, if $r \geq 3$ then*

$$f_i^2 f_j - (p^{e/2} + p^{-e/2}) f_i f_j f_i + f_j f_i^2 = 0 \quad (j = i \pm 1),$$
$$f_i f_j = f_j f_i \quad (\text{otherwise});$$

if $r = 2$ and $i \neq j$ then

$$f_i^3 f_j - (p^e+1+p^{-e}) f_i^2 f_j f_i + (p^e+1+p^{-e}) f_i f_j f_i^2 - f_j f_i^3 = 0.$$

PROOF. We may verify these equalities by explicit computation. □

DEFINITION 14.21. Let $\underline{m}$ be a multisegment. The number of the segments of residue i and length l in $\underline{m}$ is denoted by $m(i; l)$.

LEMMA 14.22. *Let $\underline{n}$ be a multisegment. Consider the segments of residue i in $\underline{n}$ and order them with respect to the length of the segments. Denote these segments by $l(i; l_1), \ldots, l(i; l_N)$, where*

$$l(i; l_k) = \boxed{x_{i,k} \quad \cdots \quad i} \qquad (l_1 \leq \cdots \leq l_N).$$

$$(x_{i,k} = i - l_k + 1)$$

Let U be a subrepresentaion of $V_{\underline{n}}$ which is isomorphic to S_i and let k_0 be the integer uniquely determined by the condition that the projection of U onto $V_{l(i;l_k)}$ is 0 if $k < k_0$ and is non-zero if $k = k_0$. Then

$$V_{\underline{n}}/U \simeq V_{\underline{n} \setminus l(i;l_{k_0}) \cup l(i-1;l_{k_0}-1)}.$$

PROOF. Since $U \simeq S_i$, we must have $U \subset \oplus_{k=1}^N V_{l(i;l_k)}$. Now each $V_{l(i;l_k)}$ has a unique subrepresentation which is isomorphic to S_i. So we identify this subrepresentation with S_i and write $S_i \subset V_{l(i;l_k)}$ by abuse of notation. To prove the claim we may delete the components $V_{l(i;l_k)}$ for which U projects to 0. Thus we may assume without loss of generality that $k_0 = 1$ and that the isomorphism between S_i and U is given by the diagonal map

$$\Delta : S_i \longrightarrow S_i^{\oplus N} \subset \bigoplus_{k=1}^N V_{l(i;l_k)}.$$

To find its cokernel, we consider the natural embedding $V_{l(i;l_1)} \subset V_{l(i;l_k)}$ for $k \geq 2$, and the isomorphism

$$\begin{array}{ccc} V_{l(i;l_1)} \oplus \cdots \oplus V_{l(i;l_N)} & \longrightarrow & V_{l(i;l_1)} \oplus \cdots \oplus V_{l(i;l_N)} \\ (x_1, \ldots, x_N) & \mapsto & (x_1, x_2 - x_1, \ldots, x_N - x_1) \end{array}.$$

Composing this isomorphism with Δ, we are reduced to the case where $U \subset V_{l(i;l_1)}$. In this case the claim is obvious. □

DEFINITION 14.23. Let $\underline{m}$ be a multisegment. For each $l \geq 1$, let $\underline{m}^{+(i;l)}$ be the multisegment obtained from $\underline{m}$ by replacing one of the segment(s) of residue $i-1$ and length $l-1$ with a segment of residue i and length l.

In the following proposition, $[[k]]_{p^e}$ means $(p^{ke} - 1)/(p^e - 1)$.

PROPOSITION 14.24. *Let $\underline{m}$ be a multisegment, and let $m(i;l)$ be as above. Then there are integers $n_1, n_2, \ldots$ such that the following equality holds in $\mathcal{H}(\mathbb{F}_{p^e})$.*

$$u_{\underline{m}} f_i = \sum_{l \geq 1} (p^e)^{n_l} [[m(i;l)+1]]_{p^e} u_{\underline{m}^{+(i;l)}}.$$

PROOF. If $u_{\underline{n}}$ appears in $u_{\underline{m}} f_i$ then we have an exact sequence

$$0 \to S_i \to V_{\underline{n}} \to V_{\underline{m}} \to 0.$$

Hence $\underline{n} = \underline{m}^{+(i;l)}$ for some (i,l) by Lemma 14.22.

Let $\underline{n} = \underline{m}^{+(i;l)}$ and recall that $U \subset V_{\underline{n}}$ means that U is a subrepresentation of $V_{\underline{n}}$. Then, up to a power of p^e, the coefficient of $u_{\underline{n}}$ in $u_{\underline{m}} f_i$ is

$$\left| \{ U \subset V_{\underline{n}} \,|\, U \simeq S_i,\, V_{\underline{n}}/U \simeq V_{\underline{m}} \} \right|.$$

We denote by $l(i;k)$ the segment of residue i and length k, and by $n(i;k)$ the multiplicity of $l(i;k)$ in $\underline{n}$. As before, $U \simeq S_i$ implies that $U \subset \oplus_{k \geq 1} V_{l(i;k)}^{\oplus n(i;k)}$. We shall consider the quotient representation $(\oplus_{k \geq 1} V_{l(i;k)}^{\oplus n(i;k)})/U$.

If $U \not\subset \oplus_{k \geq l} V_{l(i;k)}^{\oplus n(i;k)}$ then Lemma 14.22 says that $V_{\underline{n}}/U$ corresponds to the multisegment which is obtained from $\underline{n}$ by shortening a segment of length less than l by one. As this multisegment cannot be $\underline{m}$, we must have $U \subset \oplus_{k \geq l} V_{l(i;k)}^{\oplus n(i;k)}$.

On the other hand, if $U \subset \oplus_{k > l} V_{l(i;k)}^{\oplus n(i;k)}$ then the multisegment corresponding to $V_{\underline{n}}/U$ is the multisegment obtained by shortening a segment of length strictly greater than l by one. As this multisegment cannot be $\underline{m}$, we must have $U \not\subset \oplus_{k > l} V_{l(i;k)}^{\oplus n(i;k)}$.

These two conditions are not only necessary, but also sufficient: if U satisfies both $U \subset \oplus_{k \geq l} V_{l(i;k)}^{\oplus n(i;k)}$ and $U \not\subset \oplus_{k > l} V_{l(i;k)}^{\oplus n(i;k)}$ then we have $V_{\underline{n}}/U \simeq V_{\underline{m}}$. Therefore, the number of U in question is given by

$$\left| \{ U \subset \oplus_{k \geq l} V_{l(i;k)}^{\oplus n(i;k)} \,|\, U \simeq S_i \} \right| - \left| \{ U \subset \oplus_{k > l} V_{l(i;k)}^{\oplus n(i;k)} \,|\, U \simeq S_i \} \right|.$$

This number is equal to

$$\left| \{ U \subset S_i^{\oplus(\sum_{k \geq l} n(i;k))} \,|\, U \simeq S_i \} \right| - \left| \{ U \subset S_i^{\oplus(\sum_{k > l} n(i;k))} \,|\, U \simeq S_i \} \right|,$$

which is $[[\sum_{k \geq l} n(i;k)]]_{p^e} - [[\sum_{k > l} n(i;k)]]_{p^e}$. It is easy to see that this coincides with $[[n(i;l)]]_{p^e}$ up to a power of p^e. Hence the fact $n(i;l) = m(i;l)+1$ implies the result. □

LEMMA 14.25. *Let $\underline{m}, \underline{m}', \underline{m}''$ be multisegments. Then there exists a Laurent polynomial with integral coefficients*

$$F_{\underline{m},\underline{m}'}^{\underline{m}''}(v) \in \mathbb{Z}[v^{-1}]$$

such that the following equation holds for infinitely many p^e:

$$F_{\underline{m},\underline{m}'}^{\underline{m}''}(p^{-e/2}) = F_{\underline{m},\underline{m}'}^{\underline{m}''}(\mathbb{F}_{p^e}).$$

PROOF. Let $\mathbf{w}$ be the dimension vector of $\underline{m}'$ and let $\mathbf{d}$ be the dimension vector of $\underline{m}''$. We define a subvariety $X_{\underline{m},\underline{m}'}^{\underline{m}''}$ of $Gr := \prod_{i \in \mathbb{Z}/r\mathbb{Z}} Grass(w_i, d_i)$ as follows.

$$X_{\underline{m},\underline{m}'}^{\underline{m}''} = \left\{ U \in Gr \,\middle|\, x_{\underline{m}''} U \subset U,\ x_{\underline{m}''}|_U \in \mathcal{O}_{\underline{m}'},\ x_{\underline{m}''}|_{(V_{\underline{m}''}/U)} \in \mathcal{O}_{\underline{m}} \right\}$$

Then $|X^{\underline{m}''}_{\underline{m},\underline{m}'}(\mathbb{F}_{p^e})| = F^{\underline{m}''}_{\underline{m},\underline{m}'}(\mathbb{F}_{p^e})$. Let Fr be the natural Frobenius map on $X^{\underline{m}''}_{\underline{m},\underline{m}'}$. We consider the l-adic cohomology groups with compact support and apply the Grothendieck trace formula [**D**, 7.10]. Then

$$\left|X^{\underline{m}''}_{\underline{m},\underline{m}'}(\mathbb{F}_{p^e})\right| = \sum_{i=0}^{2\dim X^{\underline{m}''}_{\underline{m},\underline{m}'}} (-1)^i \mathrm{tr}(\mathrm{Fr}^{*e}, H^i_c(X^{\underline{m}''}_{\underline{m},\underline{m}'}, \overline{\mathbb{Q}}_l)).$$

A famous theorem of Deligne [**D**, 8.21] says that $H^i_c(X^{\underline{m}''}_{\underline{m},\underline{m}'}, \overline{\mathbb{Q}}_l)$ is filtered by Fr^*-stable subspaces and the eigenvalues of Fr^* on its succesive quotients are of the form "(a root of 1)$\times$(a power of $p^{ej/2}$)"for $j \leq i$. By restricting the range of e to multiples of an integer we can assume that all of the roots of unity are 1. Thus we conclude that $F^{\underline{m}''}_{\underline{m},\underline{m}'}(\mathbb{F}_{p^e}) = F^{\underline{m}''}_{\underline{m},\underline{m}'}(p^{-e/2})$ for infinitely many e for some $F^{\underline{m}''}_{\underline{m},\underline{m}'}(v) \in \mathbb{Z}[v^{-1}]$. □

DEFINITION 14.26. Let H_A be an A-free module with A-free basis $\{u_{\underline{m}}\}$ indexed by multisegments $\underline{m}$. Define an A-bilinear map $u_{\underline{m}} \otimes u_{\underline{m}'} \mapsto u_{\underline{m}} u_{\underline{m}'}$ by

$$u_{\underline{m}} u_{\underline{m}'} = \sum_{\underline{m}''} v^{\sum t_i w_{i+1} + \sum t_i w_i} F^{\underline{m}''}_{\underline{m},\underline{m}'}(v) u_{\underline{m}''}.$$

With this product, H_A is an A-algebra. This is the **Ringel-Hall algebra**.

Recall that $[k] = \frac{v^k - v^{-k}}{v - v^{-1}}$.

LEMMA 14.27. (1) *H_A is a unital associative algebra. Further, the elements f_i $(0 \leq i < r)$ satisfy the Serre relations.*
(2) *There are integers $n_1, n_2, \ldots$ such that the following equality holds in H_A.*

$$u_{\underline{m}} f_i = \sum_{l \geq 1} v^{n_l} [m(i;l) + 1] u_{\underline{m}+(i;l)}.$$

PROOF. (1) Using Lemma 14.25, that H_A is a unital associative algebra follows from Proposition 14.18 and the Serre relations among the f_i $(0 \leq i < r)$ follow from Lemma 14.20.
(2) This follows from Proposition 14.24 and Lemma 14.25. □

An important result about the Hall algebra is the following Proposition 14.28. The strategy of the proof is to use the characterization of the quantum algebra in terms of a symmetric bilinear form; this is used in [**Lusztig**] as the definition of the quantum algebra. We omit the proof; see [**cb-E**, Proposition 2.2] for the details.

PROPOSITION 14.28. *The A-subalgebra of H_A generated by the elements $f_i^{(n)}$ $(i \in \mathbb{Z}/r\mathbb{Z}, n \in \mathbb{N})$ is isomorphic to U_A^-.*

14.3. Some results from the geometric theory

We start with general terminology. See [**BBD**] or [**KW**, II.5,III.1].

DEFINITION 14.29. Let X be an algebraic variety over $\overline{\mathbb{F}}_p$. We denote by $D^b_c(X)$ the triangulated category $D^b_c(X, \mathbb{Q}_l)$ introduced in [**BBD**, 2.2.18] with coefficients extended to $\overline{\mathbb{Q}}_l$. An object $C^\cdot$ is a **perverse sheaf** if it satisfies the following conditions [**BBD**, (4.0.1),(4.0.2)].

$$\dim_{\overline{\mathbb{F}}_p} \mathrm{supp}(\mathcal{H}^i C^\cdot) \leq -i,$$
$$\dim_{\overline{\mathbb{F}}_p} \mathrm{supp}(\mathcal{H}^i DC^\cdot) \leq -i.$$

Here, the Verdier duality $C^\cdot \mapsto DC^\cdot$ is with respect to the dualizing sheaf.

It is known that the category of perverse sheaves is an abelian category.

Let $Y \subset X$ be a d-dimensional locally closed irreducible non-singular subvariety, $\mathcal{L}$ a locally constant sheaf on Y. Then there exists a unique $C^\cdot \in D_c^b(X)$ which satisfies

$$\mathcal{H}^i C^\cdot = 0 \ (i < -d), \quad \mathcal{H}^{-d} C^\cdot|_Y \simeq \mathcal{L},$$
$$\dim_{\overline{\mathbb{F}}_p} \mathrm{supp}(\mathcal{H}^i C^\cdot) < -i, \quad \dim_{\overline{\mathbb{F}}_p} \mathrm{supp}(\mathcal{H}^i DC^\cdot) < -i \ \ (i > -d).$$

This is the **intersection cohomology complex** of $(Y, \mathcal{L})$, and it is denoted by $IC(Y, \mathcal{L})$.

The intersection cohomology complexes with $\mathcal{L}$ associated with an irreducible representation of $\pi_1(Y)$ give the complete set of isomorphism classes of simple perverse sheaves; see [**BBD**, Theorem 4.3.1].

Next we consider a group action. Let G be an algebraic group and assume that X is a G-variety with G-action $a : G \times X \to X$. Let $p : G \times X \to X$ be the projection onto the second factor.

DEFINITION 14.30. A G-**equivariant sheaf** is a sheaf $\mathcal{F}$ on X which satisfies $a^*\mathcal{F} \simeq p^*\mathcal{F}$ such that the isomorphism is the identity on $\{e\} \times X$ and the following diagram induced by this isomorphism is commutative.

$$\begin{array}{ccccc} (\mathrm{id}_G \otimes a)^* a^* \mathcal{F} & \simeq & (\mathrm{id}_G \otimes a)^* p^* \mathcal{F} = p_{23}^* a^* \mathcal{F} & \simeq & p_{23}^* p^* \mathcal{F} \\ \| & & & & \| \\ (m \times \mathrm{id}_X)^* a^* \mathcal{F} & & \simeq & & (m \times \mathrm{id}_X)^* p^* \mathcal{F} \end{array}$$

In this diagram, $m : G \times G \to G$ is the product map, and $p_{23} : G \times G \times X \to G \times X$ is the projection onto the second and the third factors.

In applications we are always interested in the case where X has finitely many G-orbits. X is stratified by the closure relation of the orbits and all the sheaves we consider are locally constant on orbits.

Recall that a $\mathbb{Z}_l$-sheaf is the projective limit of a sequence of $\mathbb{Z}/l^n\mathbb{Z}$-sheaves ($n \in \mathbb{N}$), and a $\mathbb{Q}_l$-sheaf is a $\mathbb{Z}_l$-sheaf tensored with $\mathbb{Q}_l$. Hence a G-equivariant sheaf in $D_c^b(X)$ is understood as the projective limit of G-equivariant $\mathbb{Z}/l^n\mathbb{Z}$-sheaves tensored with $\overline{\mathbb{Q}}_l$. The definition of the objects in $D_c^b(X)$ [**BBD**, 2.2.18] is in the same spirit and if a perverse sheaf $C^\cdot$ satisfies a G-equivariance condition similar to the one above, with the understanding that isomorphisms are isomorphisms in $D_c^b(X)$ (quasi-isomorphisms), then $C^\cdot$ is called a G-**equivariant perverse sheaf** [**KW**, III.15]. Properties of perverse sheaves are summarized in [**Lusztig**, 8.1].

In Lemma 14.31 below we treat the case where G is connected and the stabilizer of a point is connected. An $IC(Y, \mathcal{L})$ is an irreducible G-equivariant perverse sheaf if and only if Y is a G-orbit and $\mathcal{L}$ is associated with an irreducible representation of the component group of the stabilizer of a point in Y. As a consequence, the complete set of irreducible G-equivariant perverse sheaves in this case is $\{IC(\mathcal{O}, \overline{\mathbb{Q}}_l)\}$, where $\mathcal{O}$ runs over the G-orbits.

Returning to our subject, we consider the following diagram again; however, this time, we consider this diagram over $\overline{\mathbb{F}}_p$.

$$\begin{array}{ccccccc} \mathcal{N}_T \times \mathcal{N}_W & \xleftarrow{p_1} & E'_{T,W} & \xrightarrow{p_2} & E''_{T,W} & \xrightarrow{p_3} & \mathcal{N}_V \\ (x_T, x_W) & \leftarrow & (x, \varphi, \psi) & \mapsto & (x, \varphi(W)) & \mapsto & x \end{array}$$

LEMMA 14.31. *Let* $\mathbf{d}$ *be a dimension vector. Then the complete set of irreducible* $G_{V_{\mathbf{d}}}$*-equivariant perverse sheaves on* $\mathcal{N}_{V_{\mathbf{d}}}$ *is given by* $\{IC(\mathcal{O}_{\underline{m}}, \overline{\mathbb{Q}}_l)\}$*, where* $\underline{m}$ *runs over the multisegments with dimension vector* $\mathbf{d}$.

PROOF. We prove that the stabilizer of the point $x = x_{\underline{m}}$ of $\mathcal{O}_{\underline{m}}$ is connected. Recall that $V_{\underline{m}}$ is a vector space with a basis indexed by the nodes of the segments of the multisegment $\underline{m}$. Let $W_{i,k}$ be the subspace spanned by the basis elements indexed by the nodes of residue i which appear as the first entries of segments of length k in $\underline{m}$. We denote the basis of $W_{i,k}$ by $u_{i,k}^{(j)}$ $(1 \le j \le n_{i,k})$. Then $W_{i,k}$ is a subspace of $\operatorname{Ker}(x^k) \cap V_i$ and

$$\operatorname{Ker}(x^k) \cap V_i = W_{i,k} \oplus \left(\operatorname{Ker}(x^{k-1}) \cap V_i + \operatorname{Ker}(x^k) \cap \operatorname{Im}(x) \cap V_i\right).$$

If $g \in G_V$ satisfies $gx = xg$ then g defines an element

$$g_{i,k} \in \operatorname{Hom}_{\overline{\mathbb{F}}_p}(W_{i,k}, \operatorname{Ker}(x^k) \cap V_i).$$

Let $\bar{g}_{i,k} \in \operatorname{End}(W_{i,k})$ be the composition of $g_{i,k}$ with the projection onto $W_{i,k}$. Then $\bar{g}_{i,k}$ is invertible for all i, k.

Conversely, assume that we are given a set of $g_{i,k}$ with $\bar{g}_{i,k} \in GL(W_{i,k})$. Define $g \in \operatorname{End}(V)$ by

$$g(x^l u_{i,k}^{(j)}) = x^l (g_{i,k} u_{i,k}^{(j)}) \quad (1 \le j \le n_{i,k}, 0 \le l < k).$$

Then it is easy to see that $gx = xg$. To prove that $g \in G_V$ we consider the following g-stable filtration on V_i.

$$V_i \supset V_i \cap \operatorname{Im}(x) \supset V_i \cap \operatorname{Im}(x^2) \supset \cdots.$$

Using this filtration we can represent g by a block triangular matrix and g is invertible if and only if each block diagonal part is invertible. Let N be such that $\operatorname{Ker}(x^N) = V_{\mathbf{d}}$. Then

$$V_i = (W_{i,1} \oplus \cdots \oplus W_{i,N}) \oplus (\operatorname{Im}(x) \cap V_i).$$

Write $g_{i,k;i,k'}$ for the component of g in $\operatorname{Hom}_{\overline{\mathbb{F}}_p}(W_{i,k}, W_{i,k'})$. Since

$$V_i \cap \operatorname{Im}(x^s) = \bigoplus_{k \ge k' \ge s} x^{k'} W_{i-k',k}$$

implies that

$$V_i \cap \operatorname{Im}(x^s) = V_i \cap \operatorname{Im}(x^{s+1}) \bigoplus \left(\bigoplus_{k \ge s} x^s W_{i-s,k} \right),$$

the block diagonal parts of g are given by the matrices

$$(g_{i-s,k;i-s,k'})_{k,k' \ge s} \quad (s = 1, 2, \ldots).$$

Hence, to know the block diagonal parts, it is enough to consider the action of g on $V_i / V_i \cap \operatorname{Im}(x)$. Now observe that

$$\operatorname{Ker}(x^k) \cap V_i = (W_{i,1} \oplus \cdots \oplus W_{i,k}) \oplus \left(\operatorname{Ker}(x^k) \cap \operatorname{Im}(x) \cap V_i\right).$$

This implies that the spaces $W_{i,1}\oplus\cdots\oplus W_{i,k}$ $(k=1,2,\dots)$ define a g-stable filtration on $V_i/V_i\cap \mathrm{Im}(x)$. Thus, each block diagonal component $(g_{i-s,k;i-s,k'})_{k,k'\geq s}$ is again block triangular, this time with respect to the decomposition $W_{i,1}\oplus\cdots\oplus W_{i,N}$, where the diagonal parts are given by $\bar{g}_{i-s,k}$ $(k\geq s)$. Therefore, that all of the $\bar{g}_{i,k}$ are invertible implies that $g\in G_V$.

This explicit description of the stabilizer proves that the stabilizer of a point is always connected. As a consequence, we know that the complete set of irreducible $G_{V_{\mathbf{d}}}$-equivariant perverse sheaves on $\mathcal{N}_{V_{\mathbf{d}}}$ is given by $IC(\mathcal{O}_{\underline{m}},\overline{\mathbb{Q}}_l)$, where $\underline{m}$ runs over the multisegments with dimension vector $\mathbf{d}$. □

DEFINITION 14.32. Let $\mathcal{P}_{V_{\mathbf{d}}}$ be the category whose objects are direct sums of irreducible $G_{V_{\mathbf{d}}}$-equivariant perverse sheaves on $\mathcal{N}_{V_{\mathbf{d}}}$. Set

$$\mathcal{Q}_{V_{\mathbf{d}}}=\bigoplus_{n\in\mathbb{Z}}\mathcal{P}_{V_{\mathbf{d}}}[n].$$

The direct sum of the Grothendieck groups of $\mathcal{Q}_{V_{\mathbf{d}}}$ is denoted by

$$\mathcal{H}=\bigoplus_{\mathbf{d}\in\mathbb{Z}^{\mathbf{r}}_{\geq 0}}K_0(\mathcal{Q}_{V_{\mathbf{d}}}).$$

Recall that $A=\mathbb{Z}[v,v^{-1}]$. Since $K_0(\mathcal{Q}_{V_{\mathbf{d}}})=\oplus_{n\in\mathbb{Z}}K_0(\mathcal{P}_{V_{\mathbf{d}}})[n]$, $\mathcal{H}$ becomes an A-module by $v[C^\cdot[n]]=[C^\cdot[n+1]]$. Here is an important remark.

REMARK 14.33. Assume that $\hat{X}$ is defined over a discrete valuation ring $R\subset\mathbb{C}$, with residue field of characteristic p, and that we obtain the variety X over $\overline{\mathbb{F}}_p$ and the variety $X_{\mathbb{C}}$ by base change to $\mathrm{Spec}(\overline{\mathbb{F}}_p)$ and $\mathrm{Spec}(\mathbb{C})$ respectively. Then

(1) As is explained in [**BBD**, 2.2.18], the perverse sheaves of $D^b_c(X,\mathbb{Q}_l)$ are obtained from perverse sheaves of $D^b_c(X,\mathbb{Z}_l)$ by tensoring with $\mathbb{Q}_l$.
(2) [**BBD**, Lemma 6.1.9] implies that there is an equivalence of categories between $\mathcal{Q}_{V_{\mathbf{d}}}$, defined for X, and $\mathcal{Q}_{V_{\mathbf{d}}}$, defined for $X_{\mathbb{C}}$. The pullback and pushforward functors are respected by this equivalence; see [**BBD**, 6.1.7,6.1.10].
(3) Write X_{et} for $X_{\mathbb{C}}$ with the etale topology and $X(\mathbb{C})$ for $X_{\mathbb{C}}$ with the usual topology. Then $D^b_c(X_{et},\mathbb{Z}_l)\simeq D^b_c(X(\mathbb{C}),\mathbb{Z}_l)$ by [**BBD**, 6.1.2.(B'')], where the right hand side is the derived category of bounded constructible $\mathbb{Z}_l$-sheaves in the usual sense. The pullback and pushforward functors are respected by this equivalence by [**BBD**, 6.1.2(C')]. Hence, there is an equivalence of categories between $\mathcal{Q}_{V_{\mathbf{d}}}$, defined in $D^b_c(X)$, and $\mathcal{Q}_{V_{\mathbf{d}}}$, defined in $D^b_c(X(\mathbb{C}),\overline{\mathbb{Q}}_l)$, and this equivalence respects the pullback and pushforward functors.

These facts imply that we can consider $\mathcal{Q}_{V_{\mathbf{d}}}$ in two ways: in the language of l-adic perverse sheaves on the variety X defined over $\overline{\mathbb{F}}_p$ as before, or in the language of perverse sheaves on the variety $X(\mathbb{C})$ defined over $\mathbb{C}$ with the usual topology of complex varieties and with the usual notion of derived categories. Since $\overline{\mathbb{Q}}_l$ is isomorphic to $\mathbb{C}$ as an abstract field, we may also replace the $D^b_c(X(\mathbb{C}),\overline{\mathbb{Q}}_l)$ with $D^b_c(X(\mathbb{C}),\mathbb{C})$.

LEMMA 14.34. *Let $T=V_{\mathbf{t}}, W=V_{\mathbf{w}}, V=V_{\mathbf{d}}$ and p_1,p_2,p_3 as before. Given $[A^\cdot]\in K_0(\mathcal{Q}_T)$ and $[B^\cdot]\in K_0(\mathcal{Q}_W)$, there exists a unique G_V-equivariant complex $[C^\cdot]$ such that $p_2^*[C^\cdot]=p_1^*([A^\cdot\otimes B^\cdot])$. Further, the product defined by*

$$[A^\cdot][B^\cdot]=v^{\sum t_iw_{i+1}+\sum t_iw_i}[Rp_{3!}C^\cdot],\quad v[A^\cdot]=[A^\cdot[1]]$$

makes $\mathcal{H}$ into a unital associative algebra which is isomorphic to H_A.

PROOF. To see the existence and uniqueness of $C^{\cdot}$, let $A^{\cdot}$ and $B^{\cdot}$ be intersection cohomology complexes which are G_T- and G_W-equivariant respectively. We define the action of $(h', h) \in G_T \times G_W$ on $E'_{T,W}$ by

$$(h', h)(x, \varphi, \psi) = (x, \varphi \circ h^{-1}, h' \circ \psi).$$

Then p_1 is $G_T \times G_W$-equivariant. Next we define the action of $g \in G_V$ on $E'_{T,W}$ by

$$g(x, \varphi, \psi) = (gxg^{-1}, g \circ \varphi, \psi \circ g^{-1}).$$

Then p_2 is G_V-equivariant.

Fix some $0 \to W \overset{\varphi_o}{\to} V \overset{\psi_o}{\to} T \to 0$ as before and let R be the unipotent radical of the stabilizer of $\varphi_o(W)$ in G_V and define $F = \{x \in \mathcal{N}_V | x\varphi_o(W) \subset \varphi_o(W)\}$.

For each $(x, \varphi, \psi) \in E'_{T,W}$, write $\varphi = g \circ \varphi_o$ and $\psi = \psi_o \circ g^{-1}$ for some $g \in G_V$ and define a map from $E'_{T,W}$ to $G_V \times_R F$ by $(x, \varphi, \psi) \mapsto (g, g^{-1}xg)$. Its inverse is given by $(g, x) \mapsto (gxg^{-1}, g \circ \varphi_o, \psi_o \circ g^{-1})$. Hence, $E'_{T,W} \simeq G_V \times_R F$ and $F \overset{p_1}{\to} \mathcal{N}_T \times \mathcal{N}_W$ is a vector bundle. Thus, p_1 is a smooth morphism with connected fibers of dimension, say d_1.

Let a and p be the maps $G_V \times E'_{T,W} \to E'_{T,W}$ giving the G_V-action and the projection to the second factor respectively. As $p_1 \circ a = p_1 \circ p$, $p_1^*(A^{\cdot} \otimes B^{\cdot})$ is a G_V-equivariant complex. More explicitly, since p_1 is a smooth morphism with connected fibers of dimension d_1, we may write

$$p_1^*(A^{\cdot} \otimes B^{\cdot})[d_1] = IC(p_1^{-1}(\mathcal{O}), \overline{\mathbb{Q}}_l)$$

where $\mathcal{O}$ is a $G_T \times G_W$-orbit in $\mathcal{N}_T \times \mathcal{N}_W$. On the other hand, since p_2 is a $G_T \times G_W$-principal bundle, we can write

$$p_2^*(IC(p_1^{-1}(\mathcal{O})/G_T \times G_W, \overline{\mathbb{Q}}_l))[d_2] = IC(p_1^{-1}(\mathcal{O}), \overline{\mathbb{Q}}_l),$$

in a unique way where d_2 is an integer. Thus we have the uniqueness and existence.

Now we consider the following commutative diagrams, where the horizontal arrows are equal to either a or p.

$$\begin{array}{ccc} G_V \times E''_{T,W} & \longrightarrow & E''_{T,W} \\ \downarrow & & \downarrow \\ G_V \times \mathcal{N}_V & \longrightarrow & \mathcal{N}_V \end{array}$$

Then computations like $a^*(Rp_{3!}C^{\cdot}) \simeq (\mathrm{id} \times Rp_{3!})(a^*C^{\cdot}) \simeq (\mathrm{id} \times Rp_{3!})(p^*C^{\cdot}) \simeq p^*(Rp_{3!}C^{\cdot})$ etc. show that $Rp_{3!}C^{\cdot}$ is G_V-equivariant. By [**BBD**, Corollaire 5.3.2, Remarque 5.4.9] and the decomposition theorem [**BBD**, Corollaire 5.46], $Rp_{3!}C^{\cdot}$ is isomorphic to a direct sum of shifts of simple perverse sheaves, and $Rp_{3!}C^{\cdot} \simeq \oplus_{i\in\mathbb{Z}}\mathcal{H}^i(Rp_{3!}C^{\cdot})[-i]$. Since each $\mathcal{H}^i(Rp_{3!}C^{\cdot})$ is G_V-equivariant, downward induction on the dimension of the support proves that $\mathcal{H}^i(Rp_{3!}C^{\cdot}) \in \mathcal{Q}_V$. Hence the product is well-defined.

To prove the remaining statements, we descend to finite fields. For $F^{\cdot} \in \mathcal{Q}_V$, take $\mathrm{Fr}^{e*}(F^{\cdot}) \simeq F^{\cdot}$ and define a linear map from $K_0(\mathcal{Q}_V)$ to $\mathcal{H}_V(\mathbb{F}_{p^e})$ by

$$[F^{\cdot}] \mapsto f(x) = \sum_{i\in\mathbb{Z}}(-1)^i \mathrm{tr}(\mathrm{Fr}^{e*}, \mathcal{H}^i_x(F^{\cdot})).$$

Note that the pullback p_1^* on the left hand side corresponds to the pullback p_1^* on the right hand side; namely, the pullback in the category of finite sets defined in Section 14.2, because

$$\sum_{i\in\mathbb{Z}}(-1)^i\mathrm{tr}(\mathrm{Fr}^{e*},\mathcal{H}^i_x(p_1^*F^\cdot))=\sum_{i\in\mathbb{Z}}(-1)^i\mathrm{tr}(\mathrm{Fr}^{e*},\mathcal{H}^i_{p_1(x)}(F^\cdot))=f(p_1(x)).$$

The same is true for the pullback p_2^*. The function corresponding to $C^\cdot$ is the function f'' defined in Lemma 14.13.

Similarly, since $[F^\cdot]=\sum_{j\in\mathbb{Z}}(-1)^j[\mathcal{H}^j(F^\cdot)]$ we have

$$\sum_{i\in\mathbb{Z}}(-1)^i\mathrm{tr}(\mathrm{Fr}^{e*},\mathcal{H}^i_x(Rp_{3!}F^\cdot))=\sum_{i\in\mathbb{Z}}\sum_{j\in\mathbb{Z}}(-1)^{i+j}\mathrm{tr}(\mathrm{Fr}^{e*},H^i_c(p_3^{-1}(x),\mathcal{H}^j(F^\cdot))),$$

and by the Grothendieck trace formula we get

$$\sum_{i\in\mathbb{Z}}(-1)^i\mathrm{tr}(\mathrm{Fr}^{e*},H^i_c(p_3^{-1}(x),\mathcal{H}^j(F^\cdot)))=\sum_{z\in E''(\mathbb{F}_{p^e})\cap p_3^{-1}(x)}\mathrm{tr}(\mathrm{Fr}^{e*},\mathcal{H}^j_z(F^\cdot)).$$

Thus $[Rp_{3!}F^\cdot]$ corresponds to the function $\sum_{z\in p_3^{-1}(x)}f(z)$. That is, the pushforward $Rp_{3!}$ on the left hand side corresponds to the pushforward $p_{3!}$ in the category of finite sets.

Let $\iota_{\underline{m}}$ be the embedding $\mathcal{O}_{\underline{m}}\to\mathcal{N}_V$. Then $(\iota_{\underline{m}})_!\overline{\mathbb{Q}}_l$ corresponds to the characteristic function $u_{\underline{m}}$. To summarize, we get an algebra isomorphism $\mathcal{H}\simeq\mathcal{H}_A$ and Lemma 14.27(1) implies that $\mathcal{H}$ is a unital associative algebra. □

By Remark 14.33 we may think of $\mathcal{H}$ in terms of the complex varieties $\mathcal{N}_{V_{\mathbf{d}}}$. For the rest of these notes we adopt this point of view.

To describe the A-subalgebra of $\mathcal{H}$ generated by $\{f_i^{(n)}\,|\,i\cdot\in\mathbb{Z}/r\mathbb{Z},\,n\in\mathbb{Z}\}$ geometrically, we introduce the algebra $\mathcal{K}$.

Definition 14.35. (1) Let S_i be the $\mathbb{Z}/r\mathbb{Z}$-graded vector space with dimension vector $(d_j)_{0\le j\le r-1}=(\delta_{ij})_{0\le j\le r-1}$ as before. Then the constant sheaf on $\mathcal{N}_{S_i}=\{0\}$ defines an element of $\mathcal{H}$. We denote this element by f_i.

(2) Let $\mathbf{d}\in(\mathbb{Z}/r\mathbb{Z})^r$. A **flag** F on $V_{\mathbf{d}}$ is a decreasing sequence of subspaces

$$F=(V_{\mathbf{d}}=F_0\supset F_1\supset\cdots\supset F_N=0)$$

of $V_{\mathbf{d}}$. For each pair $(\underline{i},\underline{c})$ with

$$\underline{i}=(i_1,\dots,i_N)\in(\mathbb{Z}/r\mathbb{Z})^N,\quad \underline{c}=(c_1,\dots,c_N)\in\mathbb{N}^N,$$

we consider those flags F where F_{k-1} is obtained from F_k by adding a c_{i_k}-dimensional subspace of V_{i_k} $(1\le k\le N)$. Combining such flags with the $x\in\mathcal{N}_{V_{\mathbf{d}}}$ which stabilize them gives the following variety.

$$\mathcal{F}_{\underline{i},\underline{c}}=\{(x,F)\,|\,x\in\mathcal{N}_{V_{\mathbf{d}}},\,xF_k\subset F_k\ (0\le k\le N)\,\}.$$

We pushforward the constant sheaf $\mathbb{C}$ on $\mathcal{F}_{\underline{i},\underline{c}}$ to $\mathcal{N}_{V_{\mathbf{d}}}$. As before, this is a direct sum of shifts of irreducible $G_{V_{\mathbf{d}}}$-equivariant perverse sheaves.

Let $\mathcal{P}^0_{V_{\mathbf{d}}}$ be the category whose objects are direct sums of the irreducible $G_{V_{\mathbf{d}}}$-equivariant perverse sheaves obtained in this way for various $(\underline{i},\underline{c})$. We define $\mathcal{Q}^0_{V_{\mathbf{d}}}$ by

$$\mathcal{Q}^0_{V_{\mathbf{d}}}=\bigoplus_{n\in\mathbb{Z}}\mathcal{P}^0_{V_{\mathbf{d}}}[n],$$

and set

$$\mathcal{K} = \bigoplus_{\mathbf{d}\in\mathbb{Z}^r_{\geq 0}} K_0(\mathcal{Q}^0_{V_\mathbf{d}}).$$

Note that $\mathcal{P}^0_{V_\mathbf{d}} \subset \mathcal{P}_{V_\mathbf{d}}$ and $\mathcal{Q}^0_{V_\mathbf{d}} \subset \mathcal{Q}_{V_\mathbf{d}}$.

LEMMA 14.36. *The $\{f_i\}_{i\in\mathbb{Z}/r\mathbb{Z}}$ defined above satisfy the Serre relations.*

PROOF. This follows from Lemma 14.34, but we also prove this directly. For simplicity we assume that $r \geq 3$. To compute f_0f_1, we set $T = S_0, W = S_1$ and $V = V_0 \oplus V_1$ where $V_0 = \mathbb{C}, V_1 = \mathbb{C}$. Then $\mathcal{N}_T \times \mathcal{N}_W$ is a point, and thus the pullback of the constant sheaf on $\mathcal{N}_T \times \mathcal{N}_W$ is the constant sheaf on E', which is the pullback of the constant sheaf on E''. Now for each $(x, U) \in E''$, U is unique as its dimension vector is $(0,1)$, and $xU \subset U$ for all x; so $E'' = \mathcal{N}_V \simeq \mathbb{C}$. Therefore, if we denote the constant sheaf on $\mathcal{N}_{V_\mathbf{d}}$ ($\mathbf{d} = (1,1)$) by $\mathbb{C}$, then $t_0w_1+t_0w_0+t_1w_1 = 1$ so that $f_0f_1 = \mathbb{C}[1]$.

Next we set $T = S_1, W = V_\mathbf{d}$ and $V = V_0 \oplus V_1$ where $V_0 = \mathbb{C}, V_1 = \mathbb{C}^2$ in order to compute $f_1f_0f_1$. As in the above computation, we compute the pullback of the sheaf $\mathbb{C}[1]$ on $\mathcal{N}_T \times \mathcal{N}_W \simeq \mathbb{C}$ to E'. Then we again get the sheaf $\mathbb{C}[1]$ on E''. As $t_0w_1+t_0w_0+t_1w_1 = 1$, we need to compute $Rp_{3!}\mathbb{C}[2]$.

Observe that $\mathcal{N}_V \simeq \mathbb{C}^2 =: \mathbb{A}^2$; so we write $x = (a,b)$. As the dimension condition implies that $U_0 = V_0$ and $U_1 \subset V_1$, we may and do identify U with an element of $\mathbb{P}^1$; so U is denoted by $[c:d]$. Then the condition $xU \subset U$ is nothing but $ad - bc = 0$, and we have an explicit description of p_3 as follows.

$$p_3 : \{((a,b),[c:d]) \in \mathbb{A}^2 \times \mathbb{P}^1 \mid ad - bc = 0\} \to \{(a,b) \in \mathbb{A}^2\}$$

Set $\mathcal{O}_0 = \{(0,0)\}$ and $\mathcal{O}_1 = \mathbb{A}^2 \setminus \{(0,0)\}$. Then p_3 is isomorphic to its image on $\mathcal{O}_1$, and if $x \in \mathcal{O}_0$ then its stalk is

$$(R^ip_{3!}\mathbb{C})_x = H^i_c(\mathbb{P}^1, \mathbb{C}) = \begin{cases} \mathbb{C} & (i = 0,2), \\ 0 & (\textit{otherwise}). \end{cases}$$

We denote by u_0 and u_1 the constant sheaves on $\mathcal{O}_0$ and $\mathcal{O}_1$ respectively. Recalling that $Rp_{3!}\mathbb{C} \simeq \oplus_{i\in\mathbb{Z}}\mathcal{H}^i[-i]$, we obtain

$$f_1f_0f_1 = v^2\left((1+v^{-2})u_0 + u_1\right) = (1+v^2)u_0 + v^2u_1.$$

Similar computations show that $f_0f_1^{(2)} = v^2(u_0+u_1)$ and $f_1^{(2)}f_0 = u_0$. Thus

$$f_0f_1^{(2)} - f_1f_0f_1 + f_1^{(2)}f_0 = 0.$$

We can verify other Serre relations similarly. □

Now we are in a position to state the key results obtained by Lusztig. The main point of Theorem 14.38(1) below is that $\mathcal{K}$ is no bigger than the A-subalgebra of $\mathcal{H}$ generated by $f_i^{(n)}$; see [**Lusztig**, Theorem 13.2.11].

DEFINITION 14.37. A multisegment $\underline{m}$ is **aperiodic** if the set of the last entries of the segments of length l is a proper subset of $\mathbb{Z}/r\mathbb{Z}$ for each l.

THEOREM 14.38. (1) *The A-subalgebra of $\mathcal{H}$ generated by the elements $f_i^{(n)}$ ($i \in \mathbb{Z}/r\mathbb{Z}, n \in \mathbb{N}$) coincides with $\mathcal{K}$. In particular, this A-subalgebra is a direct summand of $\mathcal{H}$ as an A-module.*

(2) *$\mathcal{K}$ is isomorphic to U_A^- as an A-algebra.*

(3) *If we set* $b_{\underline{m}} = [IC(\mathcal{O}_{\underline{m}}, \mathbb{C})]$ *then the set of* $b_{\underline{m}}$ *with aperiodic* $\underline{m}$ *is an* A*-basis of* U_A^-. *By definition this basis coincides with the canonical basis.*
(4) *Let* σ *be the anti-automorphism of* U_A^- *defined by* $\sigma(f_i) = f_i$. *Then* σ *permutes the elements of the canonical basis.*

Part (2) follows from (1), Proposition 14.28 and Lemma 14.34. Part (3) gives a precise description of the basis elements which belong to $\mathcal{K}$; see [**cb-D**, 5.4] and [**cb-D**, Theorem 5.9]. (4) is already stated in Theorem 7.3(6). The reference to the proof of this statement is given there; see also [**Lusztig**, 13.1.13].

DEFINITION 14.39. Let $q = \sqrt[r]{1} \in \mathbb{C}$ $(r \geq 2)$. To the segment of residue i and length l we associate a pair (s, x) of a diagonal matrix s and a lower triangular matrix x which satisfy $sxs^{-1} = qx$ as follows.

$$s = \begin{pmatrix} q^{i-l+1} & & & \\ & \cdot & & \\ & & \cdot & \\ & & & q^{i-1} & \\ & & & & q^{i} \end{pmatrix}, \quad x = \begin{pmatrix} 0 & & & & \\ 1 & 0 & & & \\ & \cdot & \cdot & & \\ & & \cdot & \cdot & \\ & & & 1 & 0 \end{pmatrix}$$

To a multisegment we associate the pair of block diagonal matrices obtained as the direct sum of the pairs (s, x) associated with the segments in the multisegment.

Let $\mathbf{d}$ be the dimension vector of $\underline{m}$. We denote by V_i the eigenspace of s with eigenvalue q^i. Then we may view s and x as degree 0 and degree 1 endomorphisms of $V_{\mathbf{d}}$ respectively. As $xV_i \subset V_{i+1}$, x gives the representation $V_{\underline{m}}$ of Γ.

Let $\mathcal{N}$ be the nilpotent variety, $\mathcal{B}$ the flag variety. That is, $\mathcal{N}$ consists of nilpotent endomorphisms on $V_{\mathbf{d}}$, $\mathcal{B}$ consists of decreasing sequences of subspaces $F = (F_0 \supset F_1 \supset \cdots)$ of $V_{\mathbf{d}}$ with $\dim(F_k/F_{k+1}) = 1$. Define

$$\tilde{\mathcal{N}} = \{(N, F) \in \mathcal{N} \times \mathcal{B} \mid NF_k \subset F_k \ (k = 0, 1, \ldots)\}.$$

We denote the projection to the first factor by $\pi : \tilde{\mathcal{N}} \longrightarrow \mathcal{N}$. This is the **Springer resolution** of $\mathcal{N}$.

Let (s, x) be the pair corresponding to the multisegment $\underline{m}$. We identify $\mathcal{N}_{V_{\mathbf{d}}}$ with $\{N \in \mathcal{N} \mid sNs^{-1} = qN\}$. Then $x \in \mathcal{O}_{\underline{m}} \subset \mathcal{N}_{V_{\mathbf{d}}}$. If we set

$$\tilde{\mathcal{N}}_{V_{\mathbf{d}}} = \{(N, F) \in \tilde{\mathcal{N}} \mid sNs^{-1} = qN, \ sF_k \subset F_k \ (k = 0, 1, \ldots)\},$$

and denote by $\pi_{V_{\mathbf{d}}}$ the restriction of π to $\tilde{\mathcal{N}}_{V_{\mathbf{d}}}$, then $\pi_{V_{\mathbf{d}}}^{-1}(x) = \mathcal{B}_x^s$ where $\mathcal{B}_x^s$ is the fixed points of $\mathcal{B}$ with respect to s and x. We define

$$M_{\underline{m}} = H_*(\mathcal{B}_x^s).$$

$M_{\underline{m}}$ is called a **standard module**.

Next, we write

$$\pi_{V_{\mathbf{d}}!}\mathbb{C} = \bigoplus_{\underline{m}} \bigoplus_{k \in \mathbb{Z}} L(\underline{m}, k) \otimes IC(\mathcal{O}_{\underline{m}}, \mathbb{C})[k],$$

where $\underline{m}$ is an aperiodic multisegment with $\mathcal{O}_{\underline{m}} \subset V_{\mathbf{d}}$. We define

$$L_{\underline{m}} = \bigoplus_{k \in \mathbb{Z}} L(\underline{m}, k).$$

DEFINITION 14.40. Let $q = \sqrt[r]{1} \neq 1$ be a root of unity in $\mathbb{C}$. We denote by $\hat{H}_{n,\mathbb{C}}$ the affine Hecke algebra with parameter q. Recall that $\mathcal{C}_{n,\mathbb{C}}$ is the category of $\hat{H}_{n,\mathbb{C}}$-modules introduced in Chapter 13. In the following, we denote this category by $\mathcal{C}_{n,q}$.

The following theorem of Ginzburg explains why the module $L_{\underline{m}}$ is important.

THEOREM 14.41 ([**CG**, Theorem 8.6.12]). *For each multisegment $\underline{m}$ of size n, $L_{\underline{m}}$ and $M_{\underline{m}}$ afford $\hat{H}_{n,\mathbb{C}}$-module structures. Further, we have.*

(1) *$\{L_{\underline{m}} \,|\, \underline{m}$: aperiodic of size $n\}$ is the complete set of simple objects in $\mathcal{C}_{n,\mathbb{C}}$ up to isomorphism.*
(2) *The multiplicity $[M_{\underline{m}} : L_{\underline{m}'}]$ is equal to the coefficient of $u_{\underline{m}}$ in $b_{\underline{m}'}$ evaluated at $v = 1$.*

DEFINITION 14.42. Let $\underline{m}$ be a multisegment and let $\{m_k | 1 \leq k \leq l\}$ be the set of segments in $\underline{m}$ such that m_k is of length n_k with first entry $i_k \in \mathbb{Z}/r\mathbb{Z}$. Then we denote the module $M_{(q^{i_1},\ldots,q^{i_l};n_1,\ldots,n_l)}$ by $M_{\underline{m},q}$.

Theorem 14.43 below is the Kazhdan-Lusztig induction theorem. As is stated in Chapter 12, the theorem is proved using topological K-theory. On the other hand, the definition of the standard module above comes from the K-theory (Grothendieck group) of certain equivariant sheaves on the flag variety; this is due to Ginzburg. But there is a comparison theorem. If we compare the action of the affine Hecke algebra given in [**aH-KL**, 3.10,3.11] with the action given in [**CG**, Theorem 7.2.16], then we can find the relationship between the two definitions of standard modules. So there is no harm in stating the theorem using our definition. A more detailed explanation of these facts may be found in [**cH-A2**, (3.1)-(3.8)].

THEOREM 14.43 ([**aH-KL**, Theorem 6.2]). *We have the equality*

$$[M_{\underline{m},q}] = [M_{\underline{m}}]$$

in $K_0(\mathcal{C}_{n,q})$.

Taking this theorem into account, we also call $M_{\underline{m},q}$ a **standard module**. We have considered multisegments which take values in $\mathbb{Z}/r\mathbb{Z}$. To carry out a folding argument for quantum algebras, we also need multisegments which take values in $\mathbb{Z}$. We call such a multisegment an **integral multisegment**. The **size** of an integral multisegment $\underline{\mathbf{m}}$ is the sum of the lengths of segments in $\underline{\mathbf{m}}$.

If a multisegment $\underline{m}$ is obtained from $\underline{\mathbf{m}}$ by reducing modulo r, then we write $\underline{m} = \underline{\mathbf{m}} \pmod r$.

DEFINITION 14.44. Let $\mathbf{q}$ be an indeterminate. Let $\underline{\mathbf{m}}$ be an integral multisegment and let $\{\mathbf{m}_k | 1 \leq k \leq l\}$ be the set of segments in $\underline{\mathbf{m}}$ such that $\mathbf{m}_k$ has length n_k and first entry $i_k \in \mathbb{Z}$. Then we denote the module $M_{(\mathbf{q}^{i_1},\ldots,\mathbf{q}^{i_l};n_1,\ldots,n_l)}$ by $M_{\underline{\mathbf{m}},\mathbf{q}}$. We also call $M_{\underline{\mathbf{m}},\mathbf{q}}$ a **standard module**.

Recall that $\mathcal{C}_{n,\mathbb{C}(\mathbf{q})}$ is the category of $\hat{H}_{n,\mathbb{C}(\mathbf{q})}$-modules introduced in Chapter 13. In the following, we denote this category by $\mathcal{C}_{n,\mathbf{q}}$.

If we specialize $\mathbf{q}$ to $q' \in \mathbb{C}$ which is not a root of unity, then the Kazhdan-Lusztig classification theorem [**aH-KL**, Theorem 7.12] implies that the standard modules $[M_{\underline{\mathbf{m}}}]$, where $\underline{\mathbf{m}}$ runs over the integral multisegments of size n, form a basis of the Grothendieck group $K_0(\mathcal{C}_{n,q'})$. In particular, we have the following result by specializing $\mathbf{q}$ to a complex number which is transcendental over $\mathbb{Q}$.

THEOREM 14.45. *$\{[M_{\underline{\mathbf{m}},\mathbf{q}}] \,|\, \underline{\mathbf{m}}$: integral multisegment of size $n\}$ is a basis of $K_0(\mathcal{C}_{n,\mathbf{q}})$.*

14.4. Proof of the generalized LLT conjecture

We denote by H the **specialized Hall algebra**; that is, H_A specialized at $v=1$. The $\mathbb{Q}$-subalgebra of H generated by the f_i ($i \in \mathbb{Z}/r\mathbb{Z}$) is denoted by U^-. By Theorem 14.38(1)(2), U^- is isomorphic to the enveloping algebra $U^-(\mathfrak{g}(A^{(1)}_{r-1}))$. The canonical basis of U^-_A specialized at $v=1$ is a $\mathbb{Q}$-basis of U, which we also call the **canonical basis**. We use symbols $\{b_{\underline{m}} | \,\underline{m}$: aperiodic$\}$ for the elements of the canonical basis.

Let $(\mathbb{K},\mathbb{S},\mathbb{F})$ be the modular system $(\mathbb{C}(\mathbf{q}), \mathbb{C}[\mathbf{q}]_{(\mathbf{q}-q)}, \mathbb{C})$ such that the image of $\mathbf{q}$ in the residue field $\mathbb{C}$ is q. Then we have the decomposition map $d_{\mathbb{K},\mathbb{F}}$ by Lemma 13.19; we denote this by

$$d_{\mathbf{q},q} : \mathcal{C}_{n,\mathbf{q}} \longrightarrow \mathcal{C}_{n,q}.$$

The map $d_{\mathbf{q},q}$ is surjective by Corollary 13.26.

DEFINITION 14.46. We define $U(\mathbf{q})$ and $U(q)$ by

$$U_n(\mathbf{q}) = \mathrm{Hom}_{\mathbb{Z}}(K_0(\mathcal{C}_{n,\mathbf{q}}),\mathbb{Q}), \quad U(\mathbf{q}) = \bigoplus_{n\ge 0} U_n(\mathbf{q}),$$

$$U_n(q) = \mathrm{Hom}_{\mathbb{Z}}(K_0(\mathcal{C}_{n,q}),\mathbb{Q}), \quad U(q) = \bigoplus_{n\ge 0} U_n(q).$$

The transpose of the decomposition map $d^T_{\mathbf{q},q} : U(q) \to U(\mathbf{q})$ is injective. As the standard modules give a basis of $K_0(\mathcal{C}_{n,\mathbf{q}})$ by Theorem 14.45, we can define its dual basis, which we denote by $\{[M_{\underline{\mathbf{m}},\mathbf{q}}]^* \,|\, \underline{\mathbf{m}}$: integral multisegment$\}$.

LEMMA 14.47. (1) *We define operators f_i ($i \in \mathbb{Z}/r\mathbb{Z}$) on $U(\mathbf{q})$ by*

$$f_i[M_{\underline{\mathbf{m}},\mathbf{q}}]^* = \sum_{l\ge 1} \sum_{j\in\mathbb{Z}:j\equiv i} [M_{\underline{\mathbf{m}}^{+(j;l)},\mathbf{q}}]^*,$$

where $\underline{\mathbf{m}}^{+(j;l)}$ is the integral multisegment obtained by replacing a segment of length $l-1$ in $\underline{\mathbf{m}}$ whose final entry is $j-1\in\mathbb{Z}$ by a segment of length l whose final entry is $j\in\mathbb{Z}$. Then the following diagram commutes.

$$\begin{array}{ccc} U(q) & \xrightarrow{d^T_{\mathbf{q},q}} & U(\mathbf{q}) \\ f_i \downarrow & & \downarrow f_i \\ U(q) & \xrightarrow{d^T_{\mathbf{q},q}} & U(\mathbf{q}) \end{array}$$

The operator f_i on the left hand side is $f_i = i\text{–}\mathrm{Res}^T$; see Definition 13.39.

(2) *For each multisegment $\underline{m}$ we define $e_{\underline{m}} \in U(\mathbf{q})$ by*

$$e_{\underline{m}} = \sum_{\underline{\mathbf{m}}':\underline{\mathbf{m}}' \pmod r = \underline{m}} [M_{\underline{\mathbf{m}}',\mathbf{q}}]^*.$$

These elements give a $\mathbb{Q}$-basis of a subspace of $U(\mathbf{q})$. We denote this subspace by H'. Then the operators f_i on $U(\mathbf{q})$ defined in (1) act on $e_{\underline{m}}$ by

$$f_i e_{\underline{m}} = \sum_{l\ge 1} (m(i;l)+1) e_{\underline{m}^{+(i;l)}}. \tag{14.4}$$

In particular, H' is stable under the action of f_i.

(3) *Let σ be the anti-automorphism of U^- defined by $\sigma(f_i) = f_i$. Then H becomes a U^--module by*

$$xh = h\sigma(x) \quad (x \in U^-, h \in H),$$

and the linear isomorphism $H \simeq H'$ defined by $u_{\underline{m}} \mapsto e_{\underline{m}}$ is an isomorphism of U^--modules. Under this isomorphism, we have

$$\{ b_{\underline{m}} \,|\, \underline{m} : \textit{aperiodic} \} = \{ d^T_{\mathbf{q},q}([L_{\underline{m}}]^*) \,|\, \underline{m} : \textit{aperiodic} \}, \tag{14.5}$$

where the left hand side is the canonical basis and the right hand side is the dual basis of the basis $\{[L_{\underline{m}}] \,|\, \underline{m} : \textit{aperiodic}\}$.

PROOF. (1) Let $\underline{n}$ be a multisegment. We view $\underline{n}$ as a multipartition λ whose components $\lambda^{(k)}$ are segments viewed as one row partitions. If we consider the first entries of the segments, then this defines a Hecke algebra of type $G(m', 1, n)$ for some m'. Recalling the definition of the standard module $M_{\underline{n},q}$, we may view $[M_{\underline{n},q}]$ as the pullback of the image of $[V^\lambda]$ under the quotient map from the affine Hecke algebra to the Hecke algebra. Thus Theorem 13.6(2) implies that

$$i\text{–Res}([M_{\underline{n},q}]) = \sum_{l\geq 1} \sum_{\underline{n}'} [M_{\underline{n}',q}],$$

where $\underline{n}'$ runs through all multisegments $\underline{n}'$ with $\underline{n}'^{+(i;l)} = \underline{n}$.

Let us choose an integral multisegment $\underline{\mathbf{n}}'$ for each $\underline{n}'$ with $\underline{n}'^{+(i;l)} = \underline{n}$ such that $\underline{\mathbf{n}}' \equiv \underline{n}' \pmod r$.

To show the commutativity of the diagram, take an element $x \in U(q)$ and write

$$d^T_{\mathbf{q},q}(x) = \sum c_{\underline{\mathbf{m}}}[M_{\underline{\mathbf{m}},\mathbf{q}}]^*.$$

Then for an integral multisegment $\underline{\mathbf{n}}$ with $\underline{n} = \underline{\mathbf{n}} \pmod r$, we have

$$\langle d^T_{\mathbf{q},q}(f_i x),\, [M_{\underline{\mathbf{n}},\mathbf{q}}]\rangle = \langle \sum_{\underline{\mathbf{m}}} c_{\underline{\mathbf{m}}}[M_{\underline{\mathbf{m}},\mathbf{q}}]^*,\, \sum_{l\geq 1}\sum_{\underline{n}'} [M_{\underline{\mathbf{n}}',\mathbf{q}}]\rangle = \sum_{l\geq 1}\sum_{\underline{n}'} c_{\underline{\mathbf{n}}'}.$$

Consider the segments of $\underline{\mathbf{n}}$ which are of length l and with last entry $j \equiv i$. If we delete one of these last nodes, then the resulting multisegments are in bijection with $\underline{n}'$ above. Therefore, if we denote by $\underline{\mathbf{n}}^{-(j;l)}$ the integral multisegment $\underline{\mathbf{n}}'$ with $\underline{\mathbf{n}}'^{+(j;l)} = \underline{\mathbf{n}}$, then we may choose $\{\underline{\mathbf{n}}^{-(j;l)}\}$ as the representatives $\underline{\mathbf{n}}'$ of $\underline{n}'$. So we have the formula

$$\langle d^T_{\mathbf{q},q}(f_i x),\, [M_{\underline{\mathbf{n}},\mathbf{q}}]\rangle = \sum_{l\geq 1}\sum_{j:j\equiv i} c_{\underline{\mathbf{n}}^{-(j;l)}}.$$

The right hand side is equal to $\langle \sum c_{\underline{\mathbf{m}}} f_i [M_{\underline{\mathbf{m}},\mathbf{q}}]^*,\, [M_{\underline{\mathbf{n}},\mathbf{q}}]\rangle$ by definition. Hence the diagram is commutative.

(2) By definition $\langle\, f_i e_{\underline{m}},\, [M_{\underline{\mathbf{n}},\mathbf{q}}]\,\rangle$ is equal to

$$\sum_{l\geq 1}\left(\sum_{\underline{\mathbf{m}}':\underline{\mathbf{m}}' \pmod r = \underline{m}} \; \sum_{j\in\mathbb{Z}:j\equiv i} \langle\, [M_{\underline{\mathbf{m}}'^{+(j;l)},\mathbf{q}}]^*,\, [M_{\underline{\mathbf{n}},\mathbf{q}}]\,\rangle \right).$$

Setting $\underline{n} = \underline{\mathbf{n}} \pmod r$, the sum in the paranthesis is equal to

$$\left|\{j \in \mathbb{Z} \,|\, j \equiv i,\ \underline{\mathbf{n}}^{-(j;l)} (\mathrm{mod}\ r) = \underline{m}\}\right| = \begin{cases} m(i;l)+1 & (\underline{n} = \underline{m}^{+(i;l)}) \\ 0 & (\textit{otherwise}) \end{cases}.$$

In particular, if $\mathbf{\underline{n}} \equiv \mathbf{\underline{n}}'$ (mod r) then the coefficients of $[M_{\underline{\mathbf{n}},\mathbf{q}}]^*$ and $[M_{\underline{\mathbf{n}}',\mathbf{q}}]^*$ in $f_i e_{\underline{m}}$ are the same. So

$$f_i e_{\underline{m}} = \sum_{l \geq 1} (m(i;l)+1) e_{\underline{m}+(i;l)}$$

as desired.
(3) Lemma 14.27(2) shows that we have the following equality in H.

$$u_{\underline{m}} f_i = \sum_{l \geq 1} (m(i;l)+1) u_{\underline{m}+(i;l)}.$$

Combined with the formula (14.4) in (2), we conclude that the map $u_{\underline{m}} \mapsto e_{\underline{m}}$ is an isomorphism of U^--modules. We denote the image of 1 in $\mathcal{H}'$ by the same symbol 1. As the anti-automorphism σ permutes the elements of the canonical basis, it is enough to prove that $b_{\underline{m}'}1 = d_{\mathbf{q},q}([L_{\underline{m}'}]^*)$ for the remaining statement. Now Theorem 14.41(2) implies that

$$b_{\underline{m}'} = \sum_{\underline{m}} [M_{\underline{m}} : L_{\underline{m}'}] u_{\underline{m}}.$$

As the map $u_{\underline{m}} \mapsto e_{\underline{m}}$ is an isomorphism of U^--modules, we have

$$b_{\underline{m}'}1 = \sum_{\underline{m}} [M_{\underline{m}} : L_{\underline{m}'}] e_{\underline{m}}$$

in H'. On the other hand, if $\mathbf{\underline{m}} \equiv \mathbf{\underline{n}}$ (mod r) then

$$\langle\, d^T_{\mathbf{q},q}([L_{\underline{m}'}]^*),\, M_{\underline{\mathbf{m}},\mathbf{q}} \,\rangle = \langle\, d^T_{\mathbf{q},q}([L_{\underline{m}'}]^*),\, M_{\underline{\mathbf{n}},\mathbf{q}} \,\rangle.$$

Thus $d^T_{\mathbf{q},q}([L_{\underline{m}'}]^*) \in H'$ and we may write $d^T_{\mathbf{q},q}([L_{\underline{m}'}]^*) = \sum c_{\underline{m}} e_{\underline{m}}$. The coefficient $c_{\underline{m}}$ is given by

$$c_{\underline{m}} = \langle\, d^T_{\mathbf{q},q}([L_{\underline{m}'}]^*),\, M_{\underline{\mathbf{m}},\mathbf{q}} \,\rangle = \langle\, [L_{\underline{m}'}]^*,\, M_{\underline{m},q} \,\rangle = [M_{\underline{m},q} : L_{\underline{m}'}].$$

This implies that $b_{\underline{m}'}1 = d_{\mathbf{q},q}([L_{\underline{m}'}]^*)$ by Theorem 14.43. □

COROLLARY 14.48. *Let $q = \sqrt[r]{1} \neq 1$ and consider the left regular representation of U^-. We define a linear isomorphism $U^- \simeq U(q)$ by $b_{\underline{m}} \mapsto [L_{\underline{m}}]^*$ for each aperiodic $\underline{m}$. Then this is an isomorphism of U^--modules. In particular, $U(q)$ is a cyclic U^--module and the assumptions of Proposition 13.41 are satisfied.*

PROOF. Lemma 14.47(1) implies that the injective map $d^T_{\mathbf{q},q} : U(q) \to U(\mathbf{q})$ is a homomorphism of U^--modules. Further, Lemma 14.47(3), (14.5) and Theorem 14.38(3), Theorem 14.41(1) imply that $\mathrm{Im}(d^T_{\mathbf{q},q}) = U^-1 \subset H'$. If we compose $d^T_{\mathbf{q},q}$ with the isomorphism $H' \simeq H$ then $U(q)$ is isomorphic to the right regular representation of U^- and the isomorphism is given by $[L_{\underline{m}}]^* \mapsto \sigma(b_{\underline{m}})$. Therefore, we have the desired isomorphism $U(q) \simeq U^-$ by composing this isomorphism with the isomorphism $x \mapsto \sigma(x)$ between the right regular representation and the left regular representation of U^-. □

We are now in a position to prove Theorem 12.5. We have assumed that $\mathbb{F} = \mathbb{C}$: however, the general case can be deduced from this; see [**cH-AM1**].

Recall that $V(\mathbb{C})$ is defined by

$$V(\mathbb{C}) = \bigoplus_{n \geq 0} \mathrm{Hom}_{\mathbb{Z}}(K_0(\mathcal{H}_{n,\mathbb{C}} - mod), \mathbb{Q}),$$

where the $\mathcal{H}_{n,\mathbb{C}}$ are the Hecke algebras of type $G(m,1,n)$ for $n \geq 0$ with common parameters. We denote $V(\mathbb{C})$ by $V(q)$.

THEOREM 14.49. *We assume the parameter condition*

$$q = \sqrt[r]{1} \in \mathbb{C}\ (r \geq 2), \quad v_i = q^{\gamma_i}\ (1 \leq i \leq m)$$

as above. Recall that $V(\Lambda)$ is the irreducible $\mathfrak{g}(A^{(1)}_{r-1})$-module with highest weight $\Lambda = \sum_{i=1}^{m} \Lambda_{\gamma_i}$. As in Lemma 13.35, we identify $V(\Lambda)$ with the $\mathfrak{g}(A^{(1)}_{r-1})$-submodule of $\mathcal{F}_{\gamma_m,\ldots,\gamma_1}$ generated by the empty multipartition. Then

$$\begin{array}{cccccc} d^T_{\mathbb{K},\mathbb{C}}: & V(q) & \simeq & V(\Lambda) & \subset & \mathcal{F}_{\gamma_m,\ldots,\gamma_1} \\ & [D^\mu_\mathbb{C}]^* & \mapsto & G(\mu) = \sum_{\lambda \trianglerighteq \mu} d_{\lambda,\mu}\lambda & & \end{array}.$$

Further the set of $G(\mu)$ coincides with the canonical basis of $V(\Lambda)$.

PROOF. By virtue of Corollay 14.48, we may apply Proposition 13.41. Hence $V(q) \simeq V(\Lambda)$ and $d^T_{\mathbb{K},\mathbb{C}}$ sends $[D^\mu_\mathbb{C}]^*$ to $G(\mu)$. Corollary 14.48 also implies that we have the following commutative diagram of U^--modules,

$$\begin{array}{ccc} U(q) & \simeq & U^- \\ \downarrow & & \downarrow \\ V(q) & \simeq & V(\Lambda) \end{array}$$

where the vertical arrows are both surjective, and we have the correspondence $[L_{\underline{m}}]^* \leftrightarrow b_{\underline{m}}$ in the upper row. Thus, $\{[D^\mu_\mathbb{C}]^* \,|\, \mu : D^\mu_\mathbb{C} \neq 0\}$, the set of non-zero images of the $[L_{\underline{m}}]^*$, corresponds to the canonical basis of $V(\Lambda)$ in the lower row. $\square$

CHAPTER 15

Reference guide

In this final chapter we give the references which my lectures relied on and give some instructions on how to find other results.

- For the theory of the crystal basis and the canonical basis, the lectures rely on [**Kashiwara**] and [**Lusztig**].
- If the reader would like to learn about Kac-Moody Lie algebras and quantum groups before reading the first several chapters, I recommend V. Kac [**K**] and J. C. Jantzen [**Ja**]. If the reader would like to begin by learning the basics of Lie algebras, J. E. Humphreys [**H**] is a standard text.

  There are other textbooks on quantum groups: in particular, there are books by V. Chari and A. Pressley [**CP**] and C. Kassel [**Kas**]; but these have different flavors. If the reader can read Japanese there is a textbook by Jimbo [**Ji**] which is very easy to read.
- For the modular representation theory of Hecke algebras and finite groups of Lie type in non-describing characteristic, I recommend the survey paper R. Dipper, M. Geck, G. Hiss and G. Malle [**mo-DJHM**]. It also has very good references. The exellent book M. Geck and G. Pfeiffer [**GP**] has a style and I also recommend reading this book.

  A. Mathas [**Ma**] treats the combinatorics and representation theory of Hecke algebras and q-Schur algebras of type A. This also includes a chapter about the topics of Chapter 12-14, and it may be used as a textbook to learn this subject.
- See M. Geck [**mo-G**], which is based on his talk delivered at a Bourbaki seminar, and [**GP**, 10.6]. These include a survey on A. Lascoux, B. Leclerc, and J. Y. Thibon [**cH-LLT**] and S. Ariki [**cH-A2**].
- S. Ariki [**cr-A**] is a survey on cyclotomic Hecke algebras.
- H. Barcelo and A. Ram [**cr-BR**] try to explain what combinatorial representation theory is and M. Broué, G. Malle and J. Michel [**cr-BMM**] give results of a combinatorial flavour about the modular representation theory of the finite groups of Lie type.
- In [**cH-A1**] to [**cH-Mu**] Hecke algebras associated with complex reflection groups are studied.
- As is stated in Chapter 12, the description of the crystal graph is the same as the modular branching rule. See the papers [**sM-B1**] to [**cb-R**] for this subject.
- Instead of describing the socle of the restriction, we can study when the restriction is irreducible. Such results are obtained in [**JS1**], [**JS2**] and [**JS3**].
- To understand the result of Vigneras, we need knowledge of the representation theory of algebraic groups over p-adic fields. See C. J. Bushnell

and P. Kutzko [**BK**], Vigneras' series of papers [**na-V1**] to [**na-V4**] and A. Zelevinsky [**na-Z1**].

- Other papers related to the Hecke algebras of complex reflection groups are [**S-A**] to [**S-SS**].
- The Hecke algebra of type $G(m,1,n)$ was originally introduced in the papers [**H-A**] to [**H-C**].
- The modular representation theory of the Hecke algebra of type $G(m,1,n)$ is the theme of the last several chapters. To prove Theorem 12.5, we have used the geometric construction of quantum algebras and the canonical basis. The references for this are the papers of G. Luszig [**cb-A**] to [**cb-F**]. Related subjects are treated in the papers [**cb-BLM**] to [**cb-V**].
- We have also used geometric theory of affine Hecke algebras. See [**CG**] and the papers [**aH-G1**] to [**aH-L2**]. See also [**na-Z2**].
- The theory of crystals was originally developed from the study of solvable lattice models. Good lecture notes on this subject are [**JM**] and [**CFU**]. Other related papers are listed in [**mp-H**] to [**mp-SS**].
- The papers [**qu-BCP**] to [**qu-U**] are papers on the representation theory of affine quantum algebras.
- There is another type of combinatorial representation theory using quantum algebras: this is concerned with 0-Hecke algebras. A paper by J-Y. Thibon in [**STL**], lectures on noncommutative symmetric functions, is a good survey on this subject.
- There is a vast literature by the MIT school, which is centered around S. Fomin and R. P. Stanley. These results also belong to combinatorial representation theory; since the list is already long, I omit these references.
- Finally, we list textbooks in algebraic combinatorics, which may be viewed as the classics of combinatorial representation theory. I. G. Macdonald's famous book [**M1**] is often called "the book". I also recommend [**M2**], B. Sagan [**Sa**] and R. P. Stanley [**St**]. W. Fulton [**F**] is also a good textbook.

Combinatorial representation theory is a part of Algebraic Combinatorics. Algebraic combinatorists are interested in rather special objects: Grassmannian manifolds, flag varieties, Steinberg varieties, the symmetric group, classical groups of Lie type. But we live in a very rich world in which we may treat Schubert polynomials, quantum cohomology rings of flag varieties, the Robinson-Schensted correspondence, spherical functions, representations of Hecke algebras etc. etc. by combinatorial methods.

Bibliography

We begin by listing basic references. Then we list textbooks. Papers are classified by subjects. The abbreviation of the subjects, aH, cb, cH, cr, H, JS, mo, mp, na, qu, S and sM are in alphabetal order.

1. Basic references

[Kashiwara] M. Kashiwara, *On crystal bases of the q-analogue of universal enveloping algebras*, Duke Math. J. **63** (1991), 465-516.

[Lusztig] G. Lusztig, *Introduction to Quantum Groups*, Progress in Math. **110**, Birkhäuser, 1993.

2. Books

[BBD] A. Beilinson, J. Bernstein and P. Deligne, *Faisceaux pervers*, Astérisque **100**, Soc. Math. France, 1982.

[BK] C. J. Bushnell and P. Kutzko, *The admissible dual via open compact groups*, Annals of Math. Studies **129**, Princeton University Press, 1993.

[CP] V. Chari and A. Pressley, *A Guide to Quantum Groups*, Cambridge University Press, 1994.

[CFU] I. Cherednik, P. J. Forrester and D. Uglov, *Quantum Many-Body Problems and Representation Theory*, M.S.J. Memoirs **1**, Math. Soc. Japan, 1998.

[CG] N. Chriss and V. Ginzburg, *Representation Theory and Complex Geometry*, Birkhäuser, 1997.

[D] V. I. Danilov, *Cohomology of algebraic varieties, in Algebraic geometry II*, I. R. Shafarevich eds. Encyclopaedia of Mathematical Sciences **35** (1996), 1-125.

[F] W. Fulton, *Young Tableaux, with Applications to Representation Theory and Geometry*, London Math. Soc. Student Texts **35**, Cambridge University Press, 1997.

[GP] M. Geck and G. Pfeiffer, Characters of Finite Coxeter Groups and Iwahori-Hecke Algebras, London Math. Soc. Monographs **New Series 21**, Clarendon Press Oxford, 2000.

[H] J. E. Humphreys, *Introduction to Lie Algebras and Representation Theory*, Graduate Texts in Math. **9**, Springer-Verlag, 1972.

[JK] G. James and A. Kerber, *The Representation Theory of the Symmetric Group*, Encyclopedia of Mathematics and its applications **16**, Addison-Wesley, 1981.

[Ja] J. C. Jantzen, *Lectures on Quantum Groups*, Graduate Studies in Math. **6**, Birkhäuser, 1996.

[Ji] M. Jimbo, *Quantum Groups and Yang-Baxter equations* (Japanese), Springer-Tokyo, 1990.

[JM] M. Jimbo and T. Miwa, *Algebraic Analysis of Solvable Lattice Models*, CBMS Regional Conference Series **85**, Amer. Math. Soc., 1995.

[K] V. Kac, *Infinite Dimensional Lie Algebras, 3rd.ed.*, Cambridge University Press, 1990.

[Kas] C. Kassel, *Quantum Groups*, Graduate Texts in Math. **155**, Springer-Verlag, 1995.

[KW] R. Kiehl and R. Weissauer, *Weil Conjectures, Perverse Sheaves and l'adic Fourier Transform*, Ergebnisse der Math. **42**, Springer-Verlag, 2001.

[M1] I. G. Macdonald, *Symmetric Functions and Hall Polynomials, 2nd ed.*, Oxford Mathematical Monographs, Clarendon Press, 1995.

[M2] I. G. Macdonald, *Notes on Schubert polynomials*, Publications du LACIM, LACIM Montréal, 1991.

[Ma] A. Mathas, *Iwahori-Hecke Algebras and Schur Algebras of the Symmetric Group*, University Lecture Series **15**, Amer. Math. Soc., 1999.

[Sa] B. Sagan, *The Symmetric Group: representations, combinatorial algorithms, and symmetric functions*, Wadworth and Brooks/Cole, 1991.

[St] R. P. Stanley, *Enumerative Combinatorics 2*, Cambridge University Press, 1999.

[STL] J. R. Stembridge, J-Y Thibon and M. van Leewen, *Interaction of Combinatorics and Representation Theory*, M.S.J. Memoirs **11**, Math. Soc. Japan, 2000.

3. Papers on affine Hecke algebras

[aH-G1] V. Ginzburg, *"Lagrangean"construction for representations of Hecke algebras*, Advances in Math. **63** (1987), 100-112.

[aH-G2] V. Ginzburg, *Geometric methods in representation theory of Hecke algebras and quantum groups, in Representation Theories and Algebraic Geometry*, A. Broer and A. Daigneault eds. NATO ASI series **514** (1998), 127-183.

[aH-GRV] V. Ginzburg, N. Reshetikhin and E. Vasserot, *Quantum groups and flag varieties*, Contemp. Math. **175** (1994), 101-130.

[aH-GV] V. Ginzburg and E. Vasserot, *Langlands reciprocity for affine quantum groups of type A_n*, Intern. Math. Research Notices **3** (1993), 67-85.

[aH-KL] D. Kazhdan and G. Lusztig, *Proof of the Deligne-Langlands conjecture for Hecke algebras*, Invent. Math. **87** (1987), 153-215.

[aH-L1] G. Lusztig, *Equivariant K-theory and representations of Hecke algebras*, Proc. Amer. Math. Soc. **94** (1985), 337-342.

[aH-L2] G. Lusztig, *Representations of affine Hecke algebras*, Asterisque **171-172** (1989), 73-84.

4. Lusztig's papers on the canonical basis

As the following papers are of fundamental importance to our lectures, we separate them from other papers on the canonical basis.

[cb-A] G. Lusztig, *Canonical bases arising from quantized enveloping algebras*, J. Amer. Math. Soc. **3** (1990), 447-498.

[cb-B] G. Lusztig, *Canonical basis arising from quantized enveloping algebras II*, Progr. Theor. Phys. Suppl. **102** (1990), 175-201.

[cb-C] G. Lusztig, *Quivers perverse sheaves and quantized enveloping algebras*, J. Amer. Math. Soc. **4** (1991), 365-421.

[cb-D] G. Lusztig, *Affine quivers and canonical bases*, Publ. Math. I.H.E.S. **76** (1992), 111-163.

[cb-E] G. Lusztig, *Canonical bases and Hall algebras, in Representation Theories and Algebraic Geometry*, A. Broer and A. Daigneault eds. NATO ASI series **514** (1998), 365-399.

[cb-F] G. Lusztig, *Intersection cohomology methods in representation theory*, Proc. Intern. Congr. Math. Kyoto 1990 (1991), 155-174.

5. Papers on the canonical basis and geometric representation theory

[cb-BLM] A. A. Beilinson, G. Lusztig and R. MacPherson, *A geometric setting for the quantum deformation of GL_n*, Duke Math. J. **61** (1990), 655-677.

[cb-GL1] I. Grojnowski and G. Lusztig, *On bases of irreducible representations of quantum GL_n*, Contemp. Math. **139** (1992), 167-174.

[cb-GL2] I. Grojnowski and G. Lusztig, *A comparison of bases of quantized enveloping algebras*, Contemp. Math. **153** (1993), 11-19.

[cb-KS] M. Kashiwara and Y. Saito, *Geometric construction of crystal bases*, Duke Math. J. **89** (1997), 9-36.

[cb-N1] H. Nakajima, *Instantons on ALE spaces, quiver varieties, and Kac-Moody algebras*, Duke Math. J. **76** (1994), 365-416.

[cb-N2] H. Nakajima, *Quiver varieties and Kac-Moody algebras*, Duke Math. J. **91** (1998), 515-560.

[cb-R] M. Reineke, *On the coloured graph structure of Lusztig's canonical basis*, Math. Ann. **307** (1997), 705-723.

[cb-R1] C. M. Ringel, *Hall algebras and quantum groups*, Invent. Math. **101** (1990), 583-592.

[cb-R2] C. M. Ringel, *The composition algebra of a cyclic quiver*, Proc. London Math. Soc. (3) **66** (1993), 507-537.

[cb-VV1] M. Varagnolo and E. Vasserot, *Schur duality in the toroidal setting*, Comm. Math. Phys. **182** (1996), 469-484.

[cb-VV2] M. Varagnolo and E. Vasserot, *Double-loop algebras and the Fock space*, Invent. Math. **133** (1998), 133-159.

[cb-VV3] M. Varagnolo and E. Vasserot, *On the decomposition matrices of the quantized Schur algebra*, Duke Math. J. **100** (1999), 267-297.

[cb-V] E. Vasserot, *Affine quantum groups and equivariant K-theory*, Transform. Groups **3** (1998), 269-299.

6. Papers on cyclotomic Hecke algebras

Papers on Hecke algebras of type A and type B are also listed in this section.

[cH-A1] S. Ariki, *Representation Theory of a Hecke algebra of* $G(r,p,n)$, J. Algebra **177** (1995), 164-185.

[cH-A2] S. Ariki, *On the decomposition numbers of the Hecke algebra of* $G(m,1,n)$, J. Math. Kyoto Univ. **36** (1996), 789-808.

[cH-A3] S. Ariki, *On the classification of simple modules for cyclotomic Hecke algebras of type* $G(m,1,n)$ *and Kleshchev multipartitions*, Osaka J. Math. **38** (2001), 827-837.

[cH-A4] S. Ariki, *Uno's conjecture on representation types of Hecke algebras*, to appear in the proceedings of "Algebraic Combinatorics" (held in North Carolina, 2001).

[cH-AM1] S. Ariki and A. Mathas, *The number of simple modules of the Hecke algebras of type* $G(r,1,n)$, Math. Zeit. **233** (2000), 601-623.

[cH-AM2] S. Ariki and A. Mathas, *The representation type of Hecke algebras of type* B, Adv. Math. to appear, **math.RT/0106185**.

[cH-AM3] S. Ariki and A. Mathas, *Hecke algebras with a finite number of indecomposable modules*, preprint.

[cH-BM] K. Bremke and G. Malle, *Root systems and length functions*, Geom. Dedica. **72** (1998), 83-97.

[cH-BMR] M. Broué, G. Malle and R. Rouquier, *Complex reflection groups, braid groups, Hecke algebras*, J. Reine Angew. Math. **500** (1998), 127-190.

[cH-DJ] R. Dipper and G. D. James, *Representations of Hecke algebras of type* B_n, J. Algebra **146** (1992), 454-481.

[cH-DM] R. Dipper and A. Mathas, *Morita equivalences of Ariki-Koike algebras*, Math. Zeit. to appear, **math.RT/0004014.**

[cH-DR1] J. Du and H. Rui, *Ariki-Koike algebras with semisimple bottoms*, Math. Zeit. **234** (2000), 807-830.

[cH-DR2] J. Du and H. Rui, *Specht modules and branching rules for Ariki-Koike algebras*, Comm. Alg. **29** (2001), 4701-4719.

[cH-H1] J. Hu, *A Morita equivalence theorem for Hecke algebra* $\mathcal{H}_q(D_n)$ *when* n *is even*, Manuscripta Math., to appear.

[cH-H2] J. Hu, *On simple modules of Hecke algebras of type* D_n, Sci. China Ser.A **44** (2001), 953-960.

[cH-H3] J. Hu, *Crystal basis and simple modules for Hecke algebra of type* D_n, preprint.

[cH-H4] J. Hu, *Modular representations of Hecke algebras of type* $G(p,p,n)$, preprint.

[cH-H5] J. Hu, *Crystal bases and simple modules for Hecke algebra of type* $G(p,p,n)$, preprint.

[cH-HW] J. Hu and J-p Wang, *Hecke algebras of type* D_n *at roots of unity*, J. Algebra **212** (1999), 132-160.

[cH-LLT] A. Lascoux, B. Leclerc, and J. Y. Thibon, *Hecke algebras at roots of unity and crystal bases of quantum affine algebras*, Comm. Math. Phys. **181** (1996), 205-263.

[cH-M1] G. Malle, *Unipotente Grade imprimitiver komplexer Spiegelungsgruppen*, J. Algebra **177** (1995), 768-826.

[cH-M2] G. Malle, *On the rationality and fake degrees of characters of cyclotomic algebras*, J. Math. Sci. Univ. Tokyo **6** (1999), 647-677.

[cH-M3] G. Malle, *On the generic degrees of cyclotomic algebras*, Representation Theory **4** (2000), 342-369.

[cH-MM] G. Malle and A. Mathas, *Symmetric cyclotomic Hecke algebras*, J. Algebra **205** (1998), 275-293.

[cH-Mu] G. E. Murphy, *The representations of Hecke algebras of type A_n*, J. Algebra **173** (1995), 97-121.

7. Papers on combinatorial representation theory

These are survey papers.

[cr-A] S.Ariki, *Lectures on cyclotomic Hecke algebras, in Quantum groups and Lie theory,* A. Pressley eds. London Math. Soc. Lecture Notes Series **290** (2001), 1-22.

[cr-BR] H.Barcelo and A.Ram, *Combinatorial representation theory, in New Perspectives in Algebraic Combinatorics*, M.S.R.I.Publ. **38** (1999), 23-90.

[cr-BMM] M.Broué, G.Malle and J.Michel, *Generic blocks of finite reductive groups*, Astérisque **212** (1993), 7-92.

8. Papers which introduced the Hecke algebra of type $G(m,1,n)$

[H-A] S. Ariki, *On the semisimplicity of the Hecke algebra of* $(\mathbf{Z}/\mathbf{rZ})\wr\mathfrak{S}_\mathbf{n}$, J. Algebra **169** (1994), 216-225.

[H-AK] S. Ariki and K. Koike, *A Hecke algebra of* $(\mathbf{Z}/\mathbf{rZ})\wr\mathfrak{S}_\mathbf{n}$ *and construction of its irreducible representations*, Advances in Math. **106** (1994), 216-243.

[H-BM] M. Broué and G. Malle, *Zyklotomische Heckealgebren*, Astérisque **212** (1993), 119-189.

[H-C] I. V. Cherednik, *A new interpretation of Gelfand-Tzetlin bases*, Duke Math. J. **54** (1987), 563-577.

9. Papers on the Jantzen-Seitz problem

This approach to the Jantzen-Seitz problem is closely related to our work on Hecke algebras. The authors are the same; O. Foda, B. Leclerc, M. Okado, J. Y. Thibon and A. Welsh.

[JS1] O. Foda, B. Leclerc, M. Okado, J. Y. Thibon and A. Welsh, *RSOS models and Jantzen-Seitz representations of Hecke algebras*, Lett. Math. Phys. **43** (1998), 31-42.

[JS2] O. Foda, B. Leclerc, M. Okado, J. Y. Thibon and A. Welsh, *Combinatorics of solvable lattice models, and modular representations of Hecke algebras, in Geometric Analysis and Lie Theory in Mathematics and Physics*, A. L. Carey and M. K. Murray eds. Australian Math. Soc. Lecture Series **11** (1998), 243-290.

[JS3] O. Foda, B. Leclerc, M. Okado, J. Y. Thibon and A. Welsh, *Branching functions of $A^{(1)}_{n-1}$ and Jantzen-Seitz problem for Ariki-Koike algebras*, Advances in Math. **141** (1999), 322-365.

10. Papers on the modular representation theory of Hecke algebras

These are survey papers on the modular representation theory of Hecke algebras with its applications to the groups of Lie type in mind.

[mo-DJHM] R. Dipper, M. Geck, G. Hiss and G. Malle, *Representations of Hecke algebras and finite groups of Lie type, in Algorithmic Algebra and Number Theory,* B. H. Matzat, G. M. Greuel, and G. Hiss eds. (1999), 331-378.

[mo-G] M. Geck, *Representations of Hecke algebras at roots of unity, Séminaire Bourbaki n^o* ***836***, Asterisque **252** (1998), 33-55.

11. Papers on mathematical physics

By mathematical physics we mean here the representation theoretic study of solvable lattice models.

[mp-H] T. Hayashi, *q-analogues of Clifford and Weyl algebras - spinor and oscillator representations of quantum enveloping algebras*, Comm. Math. Phys. **127** (1990), 129-144.

[mp-IJMNT] M. Idzumi, K. Iohara, M. Jimbo, T. Miwa, T. Nakashima and T. Tokihiro, *Quantum affine symmetry in vertex models*, Int. J. Mod. Phys. **A8** (1993), 1479-1511.

[mp-JMMO] M. Jimbo, K. Misra, T. Miwa and M. Okado, *Combinatorics of representations of* $U_q(\hat{sl}(n))$ *at* $q = 0$, Comm. Math. Phys. **136** (1991), 543-566.

[mp-JMO] M. Jimbo, T. Miwa and M. Okado, *Local state probabilities of solvable lattice models: an* A_{n-1} *family*, Nucl. Phys. **B 300** (1988), 74-108.

[mp-KK] S-J. Kang and M. Kashiwara, *Quantized affine algebras and crystals with core*, Comm. Math. Phys. **195** (1998), 725-740.

[mp-KKM] S-J. Kang, M. Kashiwara and K. C. Misra, *Crystal bases of Verma modules for quantum affine Lie algebras*, Composito Math. **92** (1994), 299-325.

[mp-KMN1] S-J. Kang, M. Kashiwara, K. C. Misra, T. Miwa, T. Nakashima and A. Nakayashiki, *Affine crystals and vertex models*, Int. J. Mod. Phys. **A 7** (1992), suppl.**1A**, 449-484.

[mp-KMN2] S-J. Kang, M. Kashiwara, K. C. Misra, T. Miwa, T. Nakashima and A. Nakayashiki, *Perfect crystals of quantum affine Lie algebras*, Duke Math. J. **68** (1992), suppl.**1A**, 499-607.

[mp-K1] M. Kashiwara, *Crystalizing the q-analogue of universal enveloping algebras*, Comm. Math. Phys. **133** (1990), 249-260.

[mp-K2] M. Kashiwara, *Global crystal bases of quantum groups*, Duke Math. J. **69** (1993), 455-485.

[mp-K3] M. Kashiwara, *Crystal bases of modified quantized enveloping algebras*, Duke Math. J. **73** (1994), 383-413.

[mp-K4] M. Kashiwara, *Similarity of crystal bases*, Contemp. Math. **194** (1995), 177-186.

[mp-KMPY] M. Kashiwara, T. Miwa, J. U. H. Petersen and C. M. Yung, *Perfect crystals and q-deformed Fock spaces*, Selecta Math. **New Series 2** (1996), 415-499.

[mp-KMS] M. Kashiwara, T. Miwa and E. Stern, *Decomposition of q-deformed Fock spaces*, Selecta Math. **New Series 1** (1995), 787-805.

[mp-KN] M. Kashiwara and T. Nakashima, *Crystal graphs of the q-analogue of classical Lie algebras*, J. Algebra **165** (1994), 295-345.

[mp-KKR] S. V. Kerov, A. N. Kirillov and N. Y. Reshetikhin, *Combinatorics, Bethe Ansatz, and representations of the symmetric group*, J. Soviet Math. **36** (1987), 115-128.

[mp-KR] A. N. Kirillov and N. Y. Reshetikhin, *The Bethe Ansatz and the combinatorics of Young tableaux*, J. Soviet Math. **41** (1988), 925-955.

[mp-KSS] A. N. Kirillov, A. Schilling and M. Shimozono, *A bijection between Littlewood-Richardson tableaux and rigged configurations*, Selecta Math. **New Series**, to appear.

[mp-MM] K. C. Misra and T. Miwa, *Crystal base for the basic representation of* $U_q(\hat{\mathfrak{sl}}_n)$, Comm. Math. Phys. **134** (1990), 79-88.

[mp-N] T. Nakashima, *Crystal base and a generalization of the Littlewood-Richardson rule for the classical Lie algebras*, Comm. Math. Phys. **154** (1993), 215-243.

[mp-SS] A. Schilling and M. Shimozono, *Fermionic formulas for level-restricted generalized Kostka polynomials and coset branching functions*, Comm. Math. Phys. **220** (2001), 105-164.

12. Papers on the representation theory of the algebraic groups over non-archimedian fields

[na-BZ] I. N. Bernstein and A. Zelevinsky, *Induced representations of reductive p-adic groups I*, Ann. scient. Ec. Norm. Sup. **4 série 10** (1977), 441-472.

[na-HM] R. Howe and A. Moy, *Hecke algebra isomorphisms for* $GL(n)$ *over a p-adic field*, J. Algebra **131** (1990), 388-424.

[na-R] J. D. Rogawski, *On modules over the Hecke algebra of a p-adic group*, Invent. Math. **79** (1985), 443-465.

[na-V1] M. F. Vigneras, *Banal characteristic for reductive p-adic groups*, J. Number Theory **47** (1994), 378-397.

[na-V2] M. F. Vigneras, *Sur la conjecture locale de Langlands pour $GL(n,F)$ sur F_l*, C. R. Acad. Sci. **318** (1994), 378-397.

[na-V3] M. F. Vigneras, *A propos d'une conjecture de Langlands modulaire, in Finite Reductive Groups, related structures and representations*, M. Cabanes eds. Progress in Math. **141** (1996), 415-452.

[na-V4] M. F. Vigneras, *Induced R-representations of p-adic reductive groups*, Selecta Math. **New Series 4** (1998), 549-623.

[na-Z1] A. Zelevinsky, *Induced representations of reductive p-adic groups II. On irreducible representations of $GL(n)$*, Ann. scient. Ec. Norm. Sup. **4 série 13** (1980), 165-210.

[na-Z2] A. Zelevinsky, *The p-adic analogue of the Kazhdan-Lusztig conjecture*, Functional Anal. Appl. **15** (1981), 83-92.

13. Papers on the representation theory of quantum algebras

[qu-BCP] J. Beck, V. Chari and A. Pressley, *An algebraic characterization of the affine canonical basis*, Duke Math. J. **99** (1999), 455-487.

[qu-CK] V. Chari and M. Kleber, *Symmetric functions and representations of quantum affine algebras*, **math.QA/0011161**.

[qu-CP1] V. Chari and A. Pressley, *Quantum affine algebras and affine Hecke algebras*, Pacific J. Math. **174** (1996), 295-326.

[qu-CP2] V. Chari and A. Pressley, *Quantum affine algebras at roots of unity*, Representation Theory **1** (1997), 280-328.

[qu-CP3] V. Chari and A. Pressley, *Twisted quantum affine algebras*, Comm. Math. Phys. **196** (1998), 461-476.

[qu-GW] F. M. Goodman and H. Wenzl, *Crystal bases of quantum affine algebras and affine Kazhdan-Lusztig polynomials*, Int. Math. Res. Notices **5** (1999), 251-275.

[qu-K] M. Kleber, *Finite dimensional representations of quantum affine algebras*, Dissertation, University of California Berkeley, **math.QA/9809087**.

[qu-LT1] B. Leclerc and J. Y. Thibon, *Canonical bases of q-deformed Fock spaces*, Int. Math. Res. Notices **2** (1996), 447-456.

[qu-LT2] B. Leclerc and J. Y. Thibon, *Littlewood-Richardson coefficients and Kazhdan-Lusztig polynomials, in Combinatorial Methods in Representation Theory*, Advanced Studies in Pure Math. **28** (2000), 155-220.

[qu-STU] Y. Saito, K. Takemura and D. Uglov, *Toroidal actions on level 1 modules of $U_q(\hat{sl}_n)$*, Transform. Groups **3** (1998), 75-102.

[qu-TU1] K. Takemura and D. Uglov, *Level 0 action of $U_q(\hat{\mathfrak{sl}}_n)$ on the q-deformed Fock spaces*, Comm. Math. Phys. **190** (1998), 549-583.

[qu-TU2] K. Takemura and D. Uglov, *Representations of the quantum toroidal algebra on highest weight modules of the quantum affine algebra of type $\mathfrak{gl}_N$*, Publ. R.I.M.S. **35** (1999), 407-450.

[qu-U] D. Uglov, *Canonical bases of higher-level q-deformed Fock spaces and Kazhdan-Lusztig polynomials, in Physical Combinatorics*, Progress in Math. **191** (2000), 249-299.

14. Papers on generalized q-Schur algebras

[S-A] S. Ariki, *Cyclotomic q-Schur algebras as quotients of quantum algebras*, J. Reine Angew. Math. **513** (1999), 53-69.

[S-DJM1] R. Dipper, G. D. James and A. Mathas, *The (Q,q)-Schur algebra*, Proc. London Math. Soc. (3) **77** (1998), 327-361.

[S-DJM2] R. Dipper, G. D. James and A. Mathas, *Cyclotomic q-Schur algebras*, Math. Zeit. **229** (1998), 385-416.

[S-DS] J. Du and L. Scott, *The q-Schur^2 algebra*, Trans. A.M.S. **352** (2000), 4325-4353.

[S-GL] J. Graham and G. Lehrer, *Cellular algebras*, Invent. Math. **123** (1996), 1-34.

[S-H] J. Hu, *Schur-Weyl reciprocity between quantum groups and Hecke algebras of type $G(r,1,n)$*, Math. Zeit. **238** (2001), 505-521.

[S-SS] M. Sakamoto and T. Shoji, *Schur-Weyl reciprocity for Ariki-Koike algebras*, J. Algebra **221** (1999), 293-314.

15. Papers on the symmetric group and the Mullineaux map

[sM-B1] J. Brundan, *Modular branching rules and the Mullineux map for Hecke algebras of type* A, Proc. London Math. Soc. (3) **77** (1998), 551-581.

[sM-B2] J. Brundan, *Lowering operators for* $GL(n)$ *and quantum* $GL(n)$, Proc. Symp. Pure Math. **63** (1998), 95-114.

[sM-BK] J. Brundan and A. S. Kleshchev, *Hecke-Clifford superalgebras, crystals of type* $A_{2l}^{(2)}$ *and modular branching rule for* $\hat{S}_n$, Representation Theory **5** (2001), 317-403.

[sM-K1] A. S. Kleshchev, *Branching rules for modular representations of symmetric groups I*, J. Algebra **178** (1995), 493-511.

[sM-K2] A. S. Kleshchev, *Branching rules for modular representations of symmetric groups II*, J. Reine Angew. Math. **459** (1995), 163-212.

[sM-K3] A. S. Kleshchev, *Branching rules for modular representations of symmetric groups III; some corollaries and a problem of Mullineux*, J. London Math. Soc. **54** (1996), 25-38.

[sM-K4] A. S. Kleshchev, *Branching rules for modular representations of symmetric groups IV*, J. Algebra **201** (1996), 547-572.

[sM-LT] B. Leclerc and J. Y. Thibon, *Zelevinsky's involution at roots of unity*, J. Reine Angew. Math. **513** (1999), 33-51.

Index

Titles in This Series

26 **Susumu Ariki,** Representations of quantum algebras and combinatorics of Young tableaux, 2002

25 **William T. Ross and Harold S. Shapiro,** Generalized analytic continuation, 2002

24 **Victor M. Buchstaber and Taras E. Panov,** Torus actions and their applications in topology and combinatorics, 2002

23 **Luis Barreira and Yakov B. Pesin,** Lyapunov exponents and smooth ergodic theory, 2002

22 **Yves Meyer,** Oscillating patterns in image processing and nonlinear evolution equations, 2001

21 **Bojko Bakalov and Alexander Kirillov, Jr.,** Lectures on tensor categories and modular functors, 2001

20 **Alison M. Etheridge,** An introduction to superprocesses, 2000

19 **R. A. Minlos,** Introduction to mathematical statistical physics, 2000

18 **Hiraku Nakajima,** Lectures on Hilbert schemes of points on surfaces, 1999

17 **Marcel Berger,** Riemannian geometry during the second half of the twentieth century, 2000

16 **Harish-Chandra,** Admissible invariant distributions on reductive p-adic groups (with notes by Stephen DeBacker and Paul J. Sally, Jr.), 1999

15 **Andrew Mathas,** Iwahori-Hecke algebras and Schur algebras of the symmetric group, 1999

14 **Lars Kadison,** New examples of Frobenius extensions, 1999

13 **Yakov M. Eliashberg and William P. Thurston,** Confoliations, 1998

12 **I. G. Macdonald,** Symmetric functions and orthogonal polynomials, 1998

11 **Lars Gårding,** Some points of analysis and their history, 1997

10 **Victor Kac,** Vertex algebras for beginners, Second Edition, 1998

9 **Stephen Gelbart,** Lectures on the Arthur-Selberg trace formula, 1996

8 **Bernd Sturmfels,** Gröbner bases and convex polytopes, 1996

7 **Andy R. Magid,** Lectures on differential Galois theory, 1994

6 **Dusa McDuff and Dietmar Salamon,** J-holomorphic curves and quantum cohomology, 1994

5 **V. I. Arnold,** Topological invariants of plane curves and caustics, 1994

4 **David M. Goldschmidt,** Group characters, symmetric functions, and the Hecke algebra, 1993

3 **A. N. Varchenko and P. I. Etingof,** Why the boundary of a round drop becomes a curve of order four, 1992

2 **Fritz John,** Nonlinear wave equations, formation of singularities, 1990

1 **Michael H. Freedman and Feng Luo,** Selected applications of geometry to low-dimensional topology, 1989